Statistics for Social and Behavioral Sciences

Advisors:
S.E. Fienberg
W.J. van der Linden

For further volumes:
http://www.springer.com/series/3463

Terri D. Pigott

Advances in Meta-Analysis

 Springer

Terri D. Pigott
School of Education
Loyola University Chicago
Chicago, IL, USA

ISBN 978-1-4614-2277-8   e-ISBN 978-1-4614-2278-5
DOI 10.1007/978-1-4614-2278-5
Springer New York Dordrecht Heidelberg London

Library of Congress Control Number: 2011945854

Printed on acid-free paper

Springer is part of Springer Science+Business Media (www.springer.com)

To Jenny and Alison,
who make it all worthwhile.

Acknowledgements

I am grateful to my mentors, Ingram Olkin and Betsy Jane Becker, who were the reason I have the opportunity to write this book. Larry V. Hedges has always been in the background of everything I have accomplished in my career, and I thank him and Judy for all their support.

My graduate students, Joshua Polanin and Ryan Williams, read every chapter, and, more importantly, listened to me as I worked through the details of the book. I am a better teacher and researcher because of their enthusiasm for our work together, and their endless intellectual curiosity.

My colleagues at the Campbell Collaboration and the review authors who contribute to the Collaboration's library have been the inspiration for this book. Together we all continue to strive for high quality reviews of the evidence for social programs.

My parents, Nestor and Marie Deocampo, have provided a constant supply of support and encouragement.

As any working, single mother knows, I would not be able to accomplish anything without a network of friends who can function as substitute drivers, mothers, and general ombudspersons. I am eternally grateful to the Perri family – John, Amy and Leah – for serving as our second family. More thanks are due to the Entennman-McNulty-Oswald clan, especially Judge Sheila, Craig, Erica, Carey, and Faith, for helping us do whatever is necessary to keep the household functioning. I am indebted to Magda and Kamilla for taking care of us when we needed it most. Alex Lehr served as a substitute chauffeur when I had to teach.

Finally, I thank Rick for always being the river, and Lisette Davison for helping me transform my life.

Chicago, IL, USA

Terri D. Pigott

Contents

Chapter 1
Introduction

Abstract This chapter introduces the topics that are covered in this book. The goal of the book is to provide reviewers with advanced strategies for strengthening the planning, conduct and interpretations of meta-analyses. The topics covered include planning a meta-analysis, computing power for tests in meta-analysis, handling missing data in meta-analysis, including individual level data in a traditional meta-analysis, and generalizations from a meta-analysis. Readers of this text will need to understand the basics of meta-analysis, and have access to computer programs such as Excel and SPSS. Later chapters will require more advanced computer programs such as SAS and R, and some advanced statistical theory.

1.1 Background

The past few years have seen a large increase in the use of systematic reviews in both medicine and the social sciences. The focus on evidence-based practice in many professions has spurred interest in understanding what is both known and unknown about important interventions and clinical practices. Systematic reviews have promised a transparent and replicable method for summarizing the literature to improve both policy decisions, and the design of new studies. While I believe in the potential of systematic reviews, I have also seen this potential compromised by inadequate methods and misinterpretations of results.

This book is my attempt at providing strategies for strengthening the planning, conduct and interpretation of systematic reviews that include meta-analysis. Given the amount of research that exists in medicine and the social sciences, policy-makers, researchers and consumers need ways to organize information to avoid drawing conclusions from a single study or anecdote. One way to improve the decisions made from a body of evidence is to improve the ways we synthesize research studies.

T.D. Pigott, *Advances in Meta-Analysis*, Statistics for Social and Behavioral Sciences, DOI 10.1007/978-1-4614-2278-5_1, © Springer Science+Business Media, LLC 2012

Much of the impetus for this work derives from my experience with the Campbell Collaboration, where I have served as the co-chair of the Campbell Methods group, Methods editor, and teacher of systematic research synthesis. Two different issues have inspired this book. As Rothstein (2011) has noted, there are a number of questions always asked by research reviewers. These questions include: how many studies do I need to do a meta-analysis? Should I use random effects or fixed effects models (and by the way, what are these anyway)? How much is too much heterogeneity, and what do I do about it? I would add to this list questions about how to handle missing data, what to do with more complex studies such as those that report regression coefficients, and how to draw inferences from a research synthesis. These common questions are not yet addressed clearly in the literature, and I hope that this book can provide some preliminary strategies for handling these issues.

My second motivation for writing this book is to increase the quality of the inferences we can make from a research synthesis. One way to achieve this goal is to improve both the methods used in the review, and the interpretation of those results. Anyone who has conducted a systematic review knows the effort involved. Aside from all of the decisions that a reviewer makes throughout the process, there is the inevitable question posed by the consumers of the review: what does this all mean? What decisions are warranted by the results of this review? I hope the methods discussed in this book will help research reviewers to conduct more thorough and thoughtful analyses of the data collected in a systematic review leading to a better understanding of a given literature.

The book is organized into three sections, roughly corresponding to the stages of systematic reviews as outlined by Cooper (2009). These sections are planning a meta-analysis, analyzing complex data from a meta-analysis, and interpreting meta-analysis results. Each of these sections are outlined below.

1.2 Planning a Systematic Review

One of the most important aspects of planning a systematic review involves formulating a research question. As I teach in my courses on research synthesis, the research question guides every aspect of a synthesis from data collection through reporting of results. There are three general forms of research questions that can guide a synthesis. The most common are questions about the effectiveness of a given intervention or treatment. Many of the reviews in the Cochrane and Campbell libraries are of this form: How effective is a given treatment in addressing a given condition or problem? A second type of question examines the associations between two different constructs or conditions. For example, Sirin's (2005) work examines the strength of the correlation between different measures of socio-economic status (such as mother's education level, income, or eligibility for free school lunches) and various measures of academic achievement. Another emerging area of synthesis involves synthesizing information on the specificity and sensitivity of diagnostic tests.

After refining a research question, reviewers must search and evaluate the studies considered relevant for the review. Part of the process for evaluating studies includes the development of a coding protocol, outlining the information that will be important to extract from each study. The information coded from each study will not only be used to describe the nature of the literature collected for the review, but also may help to explain variations that we find in the results of included studies. As a frequent consultant on research syntheses, I know the importance of deep knowledge of the substantive issues in a given field for both decisions on what needs to be extracted from studies in the review, and what types of analyses will be conducted.

In Chap. 3, I focus on two common issues faced by reviewers: the choice of fixed or random effects analysis, and the planning of moderator analyses for a meta-analysis. In this chapter, I argue for the use of logic models (Anderson et al. 2011) to highlight the important mechanisms that make an intervention effective, or the relationships that may exist between conditions or constructs. Logic models not only clarify the assumptions a reviewer is making about a given research area, but also help guide the data extracted from each study, and the moderator models that should be examined. Understanding the research area and planning a priori the moderators that will be tested helps avoid problems with "fishing for significance" in a meta-analysis. Researchers have paid too little attention to the number of significance tests often conducted in a typical meta-analysis, sometimes reporting on a series of single variable moderators, analogous to conducting a series of one-way ANOVAs or t-tests. These analyses not only capitalize on chance, increasing Type I error, but they also leave the reader with an incomplete picture of how moderators are confounded with each other. In Chap. 3, I advocate for the use of logic models to guide the planning of a research synthesis and meta-analysis, for carefully examining the relationships between important moderators, and for the use of meta-regression, if possible, to examine simultaneously the association of several moderators with variation in effect size.

Another common question is: How many studies do I need to conduct a meta-analysis? Though my colleagues and I have often answered "two" (Valentine et al. 2010), the more complete answer lies in understanding the power of the statistical tests in meta-analysis. I take the approach in this book that power of tests in meta-analysis like power of any statistical test needs to be computed a priori, using assumptions about the size of an important effect in a given context, and the typical sample sizes used in a given field. Again, deep substantive knowledge of a research literature is critical for a reviewer in order to make reasonable assumptions about parameters needed for power. Chapters 4, 5, 6 discuss how to compute a priori power for a meta-analysis for tests of the mean effect size, homogeneity, and moderator analyses under both fixed and random effects models. We are often concerned about power of tests in meta-analysis in order to understand the strength of the evidence we have in a given field. If we expect few studies to exist on a given intervention, we might check a priori to see how many studies are needed to find a substantive effect. If we ultimately find fewer studies than needed to detect a substantive effect, we have a more powerful argument for conducting more primary studies. For these chapters, readers need to understand basic meta-analysis, and have access to Excel or a computer program such as SPSS or R.

1.3   Analyzing Complex Data from a Meta-analysis

One problem encountered by researchers is missing data. Missing data occurs frequently in all types of data analysis, and not just a meta-analysis. Chapter 7 provides strategies for examining the sensitivity of the results of a meta-analysis to missing data. As described in this chapter, studies can be missing, or missing data can occur at the level of the effect size, or for moderators of effect size variance. Chapter 7 provides an overview of strategies used for understanding how missing data may influence the results drawn from a review.

The final chapter in this section (Chap. 8) provides background on individual participant meta-analysis, or IPD. IPD meta-analysis is a strategy for synthesizing the individual level or raw data from a set of primary studies. While it has been used widely in medicine, social scientists have not had the opportunity to use it given the difficulties in locating the individual participant level data. I provide an overview of this technique here since agencies such as the National Science Foundation and the National Institutes of Health are requiring their grantees to provide plans for data sharing. IPD meta-analysis provides the opportunity to examine how moderators are associated with effect size variance both within and between studies. Moderator analyses in meta-analysis inherently suffer from aggregation bias – the relationships we find between moderators and effect size between studies may not hold within studies. Chapter 9 provides a discussion and guidelines on the conduct of IPD meta-analysis, with an emphasis on how to combine aggregated or study-level data with individual level data.

1.4   Interpreting Results from a Meta-analysis

Chapter 9 centers on generalizations from meta-analysis. Though Chap. 9 does not provide statistical advice, it does address a concern I have about the interpretation of the results of systematic reviews. For example, the release of the synthesis on breast cancer screening in women by the US Preventive Services Task Force (US Preventive Services Task Force 2002) was widely reported and criticized since the results seemed to contradict current practice. In education, the syntheses conducted by the National Panel on Reading also fueled controversy in the field (Ehri et al. 2001), including a number of questions about what the results actually mean for practice. Chapter 9 reviews both of these meta-analyses as a way to begin a conversation about what types of actions or decisions can be justified given the nature of meta-analytic data. All researchers involved in the conduct and use of research synthesis share a commitment to providing the best evidence available to make important decisions about social policy. Providing the clearest and most accurate interpretation of research synthesis results will help us all to reach this goal.

The final chapter, Chap. 10, provides a summary of elements I consider important in a meta-analysis. The increased use of systematic reviews and meta-analysis for policy decisions needs to be accompanied by a corresponding focus on the quality of these syntheses. The final chapter provides my view of elements that will lead to both higher quality syntheses, and then to more reasoned policy decisions.

1.5   What Do Readers Need to Know to Use This Book?

Most of the topics covered in this book assume basic knowledge of meta-analysis such as is covered in the introductory texts by Borenstein et al. (2009), Cooper (2009), Higgins and Green (2011), and Lipsey and Wilson (2000). I assume, for example, that readers are familiar with the stages of a meta-analysis: problem formulation, data collection, data evaluation, data analysis, and reporting of results as outlined by Cooper (2009). I also assume an understanding of the rationale for using effect sizes. A review of the most common effect sizes and the notation used throughout the text are given in Chap. 2. In terms of data analysis, readers should know about the reasons for using weighted means for computing the mean effect, the importance of examining the heterogeneity of effect sizes, and the types of analyses (categorical and meta-regression) used to investigate models of effect size heterogeneity. I also assume that researchers conducing systematic reviews have deep knowledge of their area of interest. This knowledge of the substantive issues is critical for making choices about the kinds of analyses that should be conducted in a given area as will be demonstrated later in the text.

Later chapters of the book cover advanced topics such as missing data, and individual participant data meta-analysis. These chapters require some familiarity with matrix algebra and multi-level modeling to understand the background for the methods. However, I hope that readers without this advanced knowledge will be able to see when these methods might be useful in a meta-analysis, and will be able to contact a statistical consultant to assist in these techniques.

In terms of computer programs used to conduct meta-analysis, I assume that the reader has access to Excel, and a standard statistical computing package such as SPSS. Both of these programs can be used for most of the computations in the chapters on power analysis. Unfortunately, the more advanced techniques presented for missing data and individual participant data meta-analysis will require the use of R, a freeware statistical package, and SAS. Each technical chapter in the book includes an appendix that provides a number of computing options for calculating the models discussed. The more complex analyses may require the use of SAS, and may also be possible using the program R. Sample programs for conducting the analyses are given in the appendices to the relevant chapters.

In addition, all of the data used in the examples are given in the Data Appendix. Readers will find a brief introduction to each data set as it appears in the text, with more detail provided in the Data Appendix. The next chapter provides an overview of the notation used in the book as well as a review of the forms of effect sizes used throughout.

References

Anderson, L.M., M. Petticrew, E. Rehfuess, R. Armstrong, E. Ueffing, P. Baker, D. Francis, and P. Tugwell. 2011. Using logic models to capture complexity in systematic reviews. *Research Synthesis Methods* 2: 33–42.

Borenstein, M., L.V. Hedges, J.P.T. Higgins, and H.R. Rothstein. 2009. *Introduction to meta-analysis*. Chicester: Wiley.

Cooper, H. 2009. *Research synthesis and meta-analysis*, 4th ed. Thousand Oaks: Sage.

Ehri, L.C., S. Nunes, S. Stahl, and D. Willows. 2001. Systematic phonics instruction helps students learn to read: Evidence from the National Reading Panel's meta-analysis. *Review of Educational Research* 71: 393–448.

Higgins, J.P.T., and S. Green. 2011. *Cochrane handbook for systematic reviews of interventions*. Oxford, UK: The Cochrane Collaboration.

Lipsey, M.W., and D.B. Wilson. 2000. *Practical meta-analysis*. Thousand Oaks: Sage Publications.

Rothstein, H.R. 2011. What students want to know about meta-analysis. Paper presented at the 6th Annual Meeting of the Society for Research Synthesis Methodology, Ottawa, CA, 11 July 2011.

Sirin, S.R. 2005. Socioeconomic status and academic achievement: A meta-analytic review of research. *Review of Educational Research* 75(3): 417–453. doi:10.3102/00346543075003417.

US Preventive Services Task Force. 2002. Screening for breast cancer: Recommendations and rationale. *Annals of Internal Medicine* 137(5 Part 1): 344–346.

Valentine, J.C., T.D. Pigott, and H.R. Rothstein. 2010. How many studies do you need? A primer on statistical power in meta-analysis. *Journal of Educational and Behavioral Statistics* 35: 215–247.

Chapter 2
Review of Effect Sizes

Abstract This chapter provides an overview of the three major effect sizes that will be used in the book: the standardized mean difference, the correlation coefficient, and the log odds ratio. The notation that will be used throughout the book is also introduced.

2.1 Background

This chapter reviews the three major types of effect sizes that will be used in this text. These three general types are those used to compare the means of two continuous variables (such as the standardized mean difference), those used for the association between two measures (such as the correlation), and those used to compare the event or incidence rate in two samples (such as the odds ratio). Below I outline the general notation that will be used when talking about a generic effect size, followed by a discussion of each family of effect sizes that will be encountered in the text. For a more thorough and complete discussion of the range of effect sizes used in meta-analysis, the reader should consult any number of introductory texts (Borenstein et al. 2009; Cooper et al. 2009; Higgins and Green 2011; Lipsey and Wilson 2000).

2.2 Introduction to Notation and Basic Meta-analysis

In this section, I introduce the notation that will be used for referring to a generic effect size, and review the basic techniques for meta-analysis. I will use T_i as the effect size in the ith study where $i = 1, \dots k$, and k is the total number of studies in the sample. Note that T_i can refer to any of the three major types of effect size that are reviewed below. Also assume that each study contributes only one effect size to the data. The generic fixed-effects within-study variance of

T.D. Pigott, *Advances in Meta-Analysis*, Statistics for Social and Behavioral Sciences, DOI 10.1007/978-1-4614-2278-5_2, © Springer Science+Business Media, LLC 2012

T_i will be given by v_i; below I give the formulas for the fixed effects within-study variance of each of the three major effect sizes.

The fixed-effects weighted mean effect size, $\overline{T}_\bullet$, is written as

$$\overline{T}_\bullet = \frac{\sum\limits_{i=1}^{k} \frac{T_i}{v_i}}{\sum\limits_{i=1}^{k} \frac{1}{v_i}} = \frac{\sum\limits_{i=1}^{k} w_i T_i}{\sum\limits_{i=1}^{k} w_i} \tag{2.1}$$

where w_i is the fixed-effects inverse variance weight or $1/v_i$. The fixed-effects variance, $v_\bullet$, of the weighted mean, $\overline{T}_\bullet$, is

$$v_\bullet = \frac{1}{\sum\limits_{i=1}^{k} w_i}. \tag{2.2}$$

The 95% confidence interval for the fixed effects weighted mean effect size is given as $\overline{T}_\bullet \pm 1.96(\sqrt{v_\bullet})$.

Once we have the fixed-effects weighted mean and variance, we need to examine whether the effect sizes are homogeneous, i.e., whether they are likely to come from a single distribution of effect sizes. The homogeneity statistic, Q, is given by

$$Q = \sum\limits_{i=1}^{k} \frac{(T_i - \overline{T}_\bullet)^2}{v_i} = \sum\limits_{i=1}^{k} w_i (T_i - \overline{T}_\bullet)^2 = \sum\limits_{i=1}^{k} w_i T_i^2 - \frac{\sum\limits_{i=1}^{k} (w_i T_i)^2}{\sum\limits_{i=1}^{k} w_i}. \tag{2.3}$$

If the effect sizes are homogeneous, Q is distributed as a chi-square distribution with $k-1$ degrees of freedom.

2.3   The Random Effects Mean and Variance

As will be discussed in the next chapter, the random effects model assumes that the effect sizes in a synthesis are sampled from an unknown distribution of effect sizes that is normally distributed with mean, θ, and variance, τ^2. Our goal in a random effects analysis is to estimate the overall weighted mean and the overall variance. The weighted mean will be estimated as in (2.1), only with a weight for each study that incorporates the variance, τ^2, among effect sizes. One estimate of τ^2 is the method of moments estimator given as

$$\hat{\tau}^2 = \begin{bmatrix} \frac{Q-(k-1)}{c} & \text{if } Q \geq k-1 \\ 0 & \text{if } Q < k-1 \end{bmatrix} \tag{2.4}$$

where Q is the value of the homogeneity test for the fixed-effects model, k is the number of studies in the sample, and c is based on the fixed-effects weights,

$$c = \sum_{i=1}^{k} w_i - \frac{\sum_{i=1}^{k} w_i^2}{\sum_{i=1}^{k} w_i}. \tag{2.5}$$

The random effects variance for the ith effect size is v_i^* and is given by

$$v_i^* = v_i + \hat{\tau}^2 \tag{2.6}$$

where v_i is the fixed effects, within-study variance of the effect size, T_i. Chapter 9, on individual participant meta-analysis, will describe other methods for obtaining an estimate of the between-subjects variance, or $\hat{\tau}^2$. The random-effects weighted mean is written as $\overline{T_\bullet^*}$, and is given by

$$\overline{T_\bullet^*} = \frac{\sum_{i=1}^{k} \frac{T_i}{v_i^*}}{\sum_{i=1}^{k} \frac{1}{v_i^*}} = \frac{\sum_{i=1}^{k} w_i^* T_i}{\sum_{i=1}^{k} w_i^*} \tag{2.7}$$

with the variance of the random-effects weighted mean given by $v_\bullet^*$ below.

$$v_\bullet^* = \sum_{i=1}^{k} \frac{1}{v_i + \hat{\tau}^2} = \sum_{i=1}^{k} w_i^* \tag{2.8}$$

The 95% confidence interval for the random effects weighted mean is given by $\overline{T_\bullet^*} \pm 1.96(\sqrt{v_\bullet^*})$.

Once we have computed the random effects weighted mean and variance, we need to test the homogeneity of the effect sizes. In a random effects model, homogeneity indicates that the variance component, τ^2, is equal to 0, that is, that there is no variation between studies. The test that the variance component zero is given by

$$Q = \sum_{i=1}^{k} w_i (T_i - \bar{T}_i)^2 \tag{2.9}$$

If the test of homogeneity is statistically significant, then the estimate of τ^2 is significantly different from zero.

2.4   Common Effect Sizes Used in Examples

In this section, I introduce the effect sizes used in the examples. The three effect
sizes used in the book are the standardized mean difference, denoted as d, the
correlation coefficient, denoted as r, and the odds-ratio, denoted as OR. I describe
each of these effect sizes and their related family of effect sizes below.

2.4.1   Standardized Mean Difference

When our studies examine differences between two groups such as men and women
or a treatment and control, we use the standardized mean difference. If $\bar{X}_i$ and $\bar{Y}_i$ are
the means of the two groups, and s_X and s_Y the standard deviations for the two
groups, the standardized mean difference is given by

$$d = c(d) \frac{\bar{X}_i - \bar{Y}_i}{s_p^2} \tag{2.10}$$

where s_p^2 is the pooled standard deviation given by

$$s_p^2 = \frac{(n_X - 1)s_X^2 + (n_Y - 1)s_Y^2}{(n_X - 1) + (n_Y - 1)}, \tag{2.11}$$

where the sample sizes for each group are n_X and n_Y, and the small sample bias
correction for d, $c(d)$, is given by

$$c(d) = 1 - \frac{3}{4(n_X + n_Y) - 9}. \tag{2.12}$$

The variance of the standardized mean difference is given by

$$v_d = \frac{n_X + n_Y}{n_X n_Y} + \frac{d^2}{2(n_X + n_Y)}. \tag{2.13}$$

The standardized mean difference, d, is the most common form of the effect size
when the studies focus on estimating differences among two independent groups
such as a treatment and a control group, or between boys and girls. Note that in the
case of the standardized mean difference, d, we assume that the unit of analysis is
the individual, not a cluster or a group.

2.4.2   Correlation Coefficient

When we are interested in the association between two measures, we use the
correlation coefficient as the effect size, denoted by r. However, the correlation

coefficient, r, is not normally distributed, and thus we use Fisher's z-transformation for our analyses. Fisher's z-transformation is given by

$$z = .5 \ln\left[\frac{1+r}{1-r}\right].$$
(2.14)

The variance for Fisher's z is

$$v_z = \frac{1}{n-3}$$
(2.15)

where n is the sample size in the study. After computing the mean correlation and its confidence interval in the Fisher's z metric, the results can be transformed back into a correlation using

$$r = \frac{e^{2z}-1}{e^{2z}+1}$$
(2.16)

The correlation coefficient, r, and Fisher's z are typically used when synthesizing observational studies, when the research question is concerned with estimating the strength of the relationship between two measures. In Chap. 9, I also provide an analysis using the raw correlation, r, rather than Fisher's z.

2.4.3   Log Odds Ratio

When we are interested in differences in incidence rates between two groups, such as comparing the number of cases of a disease in men and women, we can use a number of effect sizes such as relative risk or the odds ratio. In this book, we will use the odds ratio, OR, and its log transformation, LOR. While there are a number of effect sizes used to synthesize incidence rates or counts, here we will focus on the log odds ratio since it has desirable statistical properties (Lipsey and Wilson 2000). To illustrate the odds ratio, imagine we have data in a 2×2 table as displayed in Table 2.1.

Table 2.1 Example of data for a log odds ratio

	Group A	Group B
Condition present	a	b
Condition not present	c	d

The odds ratio for the data above is given by

$$OR = \frac{ad}{bc}$$
(2.17)

with the log-odds ratio given by

$$LOR = \ln(OR).$$

The variance of the log-odds ratio, LOR, is given by

$$v_{LOR} = \frac{1}{a} + \frac{1}{b} + \frac{1}{c} + \frac{1}{d}. \tag{2.18}$$

We use the log odds ratio, LOR, since the odds ratio, OR, has the undesirable property of being centered at 1, and with a range from 0 to ∞. The log odds ratio, LOR, is centered at 0, and ranges from $-\infty$ to ∞. The reader interested in other types of effect sizes and in more details about the distributions of these effect sizes should examine any number of texts of meta-analysis (Borenstein et al. 2009; Cooper 2009; Cooper et al. 2009; Higgins and Green 2011; Lipsey and Wilson 2000).

References

Borenstein, M., L.V. Hedges, J.P.T. Higgins, and H.R. Rothstein. 2009. *Introduction to meta-analysis*. Chicester/West Sussex/United Kingdom: Wiley.
Cooper, H. 2009. *Research synthesis and meta-analysis*, 4th ed. Thousand Oaks: Sage.
Cooper, H., L.V. Hedges, and J.C. Valentine (eds.). 2009. *The handbook of research synthesis and meta-analysis*. New York: Russell Sage Foundation.
Higgins, J.P.T., and S. Green. 2011. *Cochrane handbook for systematic reviews of interventions*. Oxford, UK: The Cochrane Collaboration.
Lipsey, M.W., and D.B. Wilson. 2000. *Practical meta-analysis*. Thousand Oaks: Sage Publications.

Chapter 3
Planning a Meta-analysis in a Systematic Review

Abstract This chapter provides guidance on planning a meta-analysis. The topics covered include choosing moderators for effect size models, considerations for choosing between fixed and random effects models, issues in conducting moderator models in meta-analysis such as confounding of predictors, and computing meta-regression. Examples are provided using data from a meta-analysis by Sirin (2005). The chapter's appendix also provides SPSS and SAS program code for the analyses in the examples.

3.1 Background

Many reviewers have difficulties in planning and estimating meta-analyses as part of a systematic review. There are a number of stages in planning and executing a meta-analysis including: (1) deciding on what information should be extracted from a study that may be used for the meta-analysis, (2) choosing among fixed, random or mixed models for the analysis, (3) exploring possible confounding of moderators in the analyses, (4) conducting the analyses, (5) interpreting the results. Each of these steps is interrelated, and all depend on the scope and nature of the research question for the review. Like any data analysis project, a meta-analysis, even if it is considered a small one, provides complex data that the researcher needs to interpret. Thus, while the literature retrieval and coding phases may take a large proportion of the time needed to complete a systematic review, the data analysis stage requires some careful thought about how to examine the data and understand the patterns that may exist. This chapter reviews the steps for conducting a moderator analysis, and provides some recommendations for practice. The justification for many of the recommendations here is best practice statistical methods; there have been many instances in the meta-analytic literature where the analytic procedures have not followed standard statistical analysis practices. If we want our research syntheses to have influence on practice, we need to make sure our results are conducted to the highest standard.

T.D. Pigott, *Advances in Meta-Analysis*, Statistics for Social and Behavioral Sciences, DOI 10.1007/978-1-4614-2278-5_3, © Springer Science+Business Media, LLC 2012

3.2   Deciding on Important Moderators of Effect Size

As many other texts on meta-analysis have noted (Cooper et al. 2009; Lipsey and
Wilson 2000), one critical stage of a research synthesis is coding of the studies. It is
in this stage that the research synthesist should have identified important aspects
of studies that need to be considered when interpreting the effects of an intervention
or the magnitude of a relationship across studies. One strategy for identifying
important aspects of studies is to develop a logic model (Anderson et al. 2011).
A logic model outlines how an intervention should work, and how different
constructs are related to one another. Logic models can be used as a blueprint to
guide the research synthesis; if the effect sizes in a review are heterogeneous, the
logic model suggests what moderator analyses should be conducted, a priori, to
avoid fishing for statistical significance in the data. Part of the logic model may be
suggested by prior claims made in the literature, i.e., that an intervention is most
effective for a particular subset of students. These claims also guide the choice of
moderator analyses. Figure 3.1 is taken from the Barel et al. (2010) meta-analysis of
the long-term sequelae of surviving genocide. As seen in the Figure, prior research
suggests that survivors' adjustment relates to the age, gender, country of residence,
and type of sample (clinical versus non-clinical). In addition, research design
quality is assumed related to the results of studies examining survivors' adjustment.

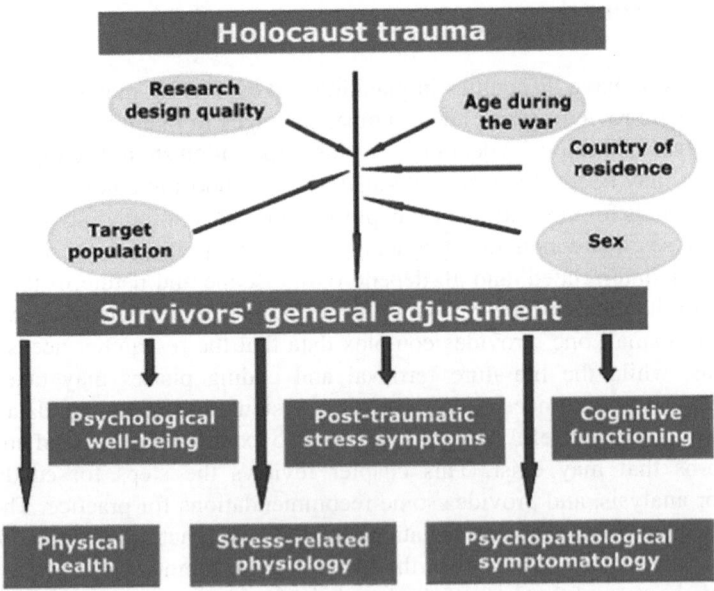

Fig. 3.1 Logic model from Barel et al. (2010)

The model shown in Fig. 3.1 also indicates the range of outcome measures included in the literature. This conceptual model guides not only the types of codes to use in the data collection phase of the synthesis, but also the analyses that will add to our understanding of these processes. Identifying a set of moderator analyses a priori that are tied to a conceptual framework avoids looking for relationships after obtaining the data, thus, capitalizing on chance to discover spurious findings. In medical research, a number of researchers have also discussed the importance of the use of logic models and causal diagrams in examining research findings (Greenland et al. 1999; Joffe and Mindell 2006).

Raudenbush (1983) illustrates another method for generating a priori ideas about possible moderator analyses. In his chapter for the Evaluation Studies Review Annual, Raudenbush outlines the controversies surrounding studies of teacher expectancy's effects on pupil IQ that were debated in the 1970s and 1980s. Researchers used the same literature base to argue both for and against the existence of a strong effect of teacher expectancy on students' measured IQ. These studies typically induced a teacher's expectancy about a student's potential by informing these teachers about how a random sample of students was expected to make large gains in their ability during the current school year. Raudenbush describes how a careful reading of the original Pygmalion study (Rosenthal and Jacobson 1968) generated a number of ideas about moderator analyses. For example, the critics of both the original and replication studies generated a number of interesting hypotheses. One of these hypotheses grew out of the failure of subsequent studies to replicate the original findings – that the timing of the expectancy induction may be important. If teachers are provided the expectancy induction after they have gotten to know the children in their class, any information that does not conform to their own assessment of the child's abilities may be discounted, leading to a smaller effect of the expectancy induction. Raudenbush illustrates how the timing of the induction does relate to the size of the effect found in these studies. An understanding of a particular literature, and as Raudenbush emphasizes, controversies in that literature can guide the research reviewer in planning a priori moderator analyses in a research synthesis.

Other important moderator analyses might be suggested by prior work in research synthesis as detailed by Lipsey (2009). In intervention research, the nature of the randomization used often relates to the size of the effect; randomized controlled trials often result in effect size estimates that are different from studies using quasi-experimental techniques. Much research has been conducted on the difference between published and unpublished studies, given the tendency in the published literature to favor statistically significant results. Littell et al. (2005) have also found that studies where the program's developer acts as the researcher/ evaluator result in larger effects for the program than studies using independent evaluators. No matter how moderators may be chosen, like in any statistical analysis, a reviewer should have a priori theories about the relationships of interest that will guide any moderator analysis.

3.3   Choosing Among Fixed, Random and Mixed Effects Models

Many introductory texts on meta-analysis have discussed this issue in depth such as Borenstein et al. (2009), as well as the Cochrane Handbook (Higgins and Green 2011). My aim here is not to repeat these thorough summaries, but to clarify what each of these choices means for the statistical analysis in a research review. These terms have been used in various ways in both the medical and social science meta-analysis literature.

The terms fixed, random and mixed effects all refer to choices that a meta-analyst has in deciding on a model for a meta-analysis. In order to clarify how these terms are used, we need to describe the type of model we are looking at in meta-analysis as well as the assumptions we are making about the error variance in the model. The first stage in a meta-analysis is usually to estimate the mean effect size and its variance, and also to examine the amount of heterogeneity that exists across studies. It is in this stage of estimating a mean effect size that the analyst needs to make a decision about the nature of the variance that exists among studies in their effect size estimates.

If we estimate the mean effect size using the fixed effects assumption, we are assuming that the variation among effect sizes can be explained by sampling error alone – that the fact that different samples are used in each study accounts for the differences in effect size magnitude. The heterogeneity among effect sizes is entirely due to the fact that the studies use different samples of subjects. Hedges and Vevea (1998) emphasize that in assuming fixed effects, the analyst wishes to make inferences only about the studies that are gathered for the synthesis. The studies in a fixed effects model are not representative of a population of studies, and are not assumed to be a random sample of studies.

If we estimate the mean effect size using the random effects assumption, we are making a two stage sampling assumption as discussed by Raudenbush (2009). We first assume that each study's effect size is a random draw from some underlying population of effect sizes. This population of effect sizes has a mean θ, and variance, τ^2. Thus, one component of variation among effect size estimates is τ^2, the variation among studies. Within each study, we use a different sample of individuals so our estimate of each study's effect size will vary from its study mean by the sampling variance, v_i. The variance among effect size estimates in a random effects model consists of within-study sampling variance, v_i, and between-study variance, τ^2. Hedges and Vevea (1998) note that when using a random effects model, we are assuming that our studies are sampled from an underlying population of studies, and that our results will generalize to this population. We are assuming in a random effects model that we have carefully sampled our studies from the literature, including as many as fit our a priori inclusion criterion, and including studies from published and unpublished sources to represent the range of possible results.

When our effect sizes are heterogeneous, and we want to explore reasons for this variation among effect size estimates, we make assumptions about whether this variation is fixed, random, or a mix of fixed and random. These choices apply to both categorical models of effect size (one-way ANOVAs, for example), or to regression models. With fixed effects models of effect size with moderators, we assume that the differences among studies can be explained by sampling error, i.e., the differences among studies in their samples and their procedures. In the early history of meta-analysis, the fixed effects assumption was the most common. More recently, reviewers have tended to use random effects models since there are multiple sources of variation among studies that reviewers do not want to attribute only to sampling variation. Random effects models also provide estimates with larger confidence levels (larger variances) since we assume a component of between study, random variance.

The confusion over mixed and fully random effects models occurs when we are talking about effect size models with random components. The most common use of the term "mixed" model refers to the hierarchical linear model formulation of meta-analysis. At one stage, each study's effect size estimate is assumed sampled from a normal distribution with mean θ, and variance, τ^2. At the level of the study, we sample individuals into the study. The two components of variation between study effect sizes are then τ^2 and v_i, the sampling variance of the study effect size. This is our typical random effects model specification. We assume that some proportion of the variation among studies can be accounted for by differences in study characteristics (i.e., sampling variance), and some is due to the underlying distribution of effect sizes.

Raudenbush (2009) also calls this mixed model variation the conditional random effect variance, conditional on the fixed moderators that represent differences in study characteristics. What is left over after accounting for fixed differences among studies in their procedures, methods, sample, etc. is the random effects variation. τ^2. Thus, some of the differences among studies may be due to fixed moderator effects, and some due to unknown random variation. For example, when we have groups such as girls versus boys, we might assume that the grouping variable or factor is fixed – that the groups in our model are not sampled from some universe of groups. Gender is usually considered a fixed factor since it has a finite number of levels. When we assume random variation within each group, but consider the levels of the factor as fixed, we have a mixed categorical model.

If, however, we consider the levels of the factor as random, such as we might do if we have sampled schools from the population of districts in a state, then we have a fully random effects model. We consider the effect sizes within groups as sampled from a population of effect sizes, and the levels of factor (the groups) as also randomly sampled from a population of levels for that factor.

Borenstein et al. (2009) provide a useful and clear description of random and mixed effects models in the context of categorical models of effect size. As they point out, if we are estimating a random effects categorical model of effect size, we also have to make some assumptions about the nature of the random variance. We can make three different choices about the nature of the random variance among the groups in our categorical model. The simplest assumption is that the random variance component is the same within each of our groups, and thus we estimate the random variance assuming that the groups differ in their true mean effect size, but have the same variance component. A second assumption is that the variance components within each group differ; one group might be assumed to have more underlying variability than another. In this scenario, we estimate the variance component within each group as a function of the differences among the effect size estimates and their corresponding group mean. For example, we might have reason to suspect that an intervention group has a larger variance after the treatment than the control group. We might also assume that each group has a separate variance component, and the differences among the means are also random. This assumption, rare in meta-analysis, is a fully random model.

All of the examples in this book will assume a common variance component among studies. Borenstein et al. (2009) discuss the difficulties in estimating the variance component with small sample sizes. Chapter 2 provided one estimate for the variance component, the method of moments, also called the DerSimonian nd Laird estimator (Dersimonian and Laird 1986). Below I illustrate another estimate of the variance component that requires iterative methods of estimation that are available in SAS or R.

3.4 Computing the Variance Component in Random and Mixed Models

The most difficult part of computing random and mixed effects models is the estimation of the variance component. As outlined by Raudenbush, to compute the random effects mean and variance in a simple random effects model requires two steps. The first step is to compute the random effects variance, and the second uses that variance estimate to compute the random effects mean. There are at least three methods for computing the random effect variance: (1) the method of moments (Dersimonian and Laird 1986), (2) full maximum likelihood, and (3) restricted maximum likelihood. Only the method of moments provides a closed solution for the random effects variance. This estimate, though easy to compute, is not efficient given that it is not based on assumptions about the likelihood. Both the full maximum likelihood and the restricted maximum likelihood solutions require iterative solutions. Fortunately, several common computing packages will provide estimates of the variance component using maximum likelihood methods. Below is an outline of how to obtain the variance component using two of these methods, the method of moments, and restricted maximum likelihood, in a simple random effects model with no moderator variables. Raudenbush (2009) compares the performance of both full maximum likelihood and restricted maximum likelihood methods, concluding that the restricted maximum likelihood method (REML) provides better estimates. The Appendix provides examples of programs used to compute the variance components using these methods.

Method of moments. This estimate can be obtained using any program that can provide the sums of various quantities of interest. SPSS Descriptives options allow the computation of the sum, and Excel can also be used. The closed form solution for the variance component τ^2 is given by Raudenbush (2009) as

$$\tau^2 = \frac{\sum_{i=1}^{k}(T_i - \bar{T})^2}{k-1} - \bar{v}$$

where $\bar{T} = \sum_{i=1}^{k} T_i/k$, and $\bar{v} = \sum_{i=1}^{k} v_i/k$. Raudenbush calls this estimate the method of moments using ordinary least squares regression – all studies are weighted the same in computing the variance component.

A more common method for computing the variance component is what Raudenbush calls the method of moments using weighted least squares. This method of computing the variance component is included in several computer programs that compute meta-analysis models such as RevMan (Cochrane Information Management System 2011) and CMA (Comprehensive Meta-Analysis Version 2 2006). The solution given below is equivalent to the estimator described in (2.3) in Chap. 2. Raudenbush gives the closed form solution of the methods of moments estimator using weighted least squares as

$$\hat{\tau}^2 = \frac{\sum_{i=1}^{k}\left[1/v_i(T_i - \bar{T})^2\right] - (k-1)}{tr(M)}, \text{ where}$$

$$tr(M) = \sum_{i=1}^{k} 1/v_i - \frac{\sum_{i=1}^{k} 1/v_i^2}{\sum_{i=1}^{k} 1/v_i}, \text{ and}$$

$$\bar{T} = \frac{\sum_{i=1}^{k} 1/v_i T_i}{\sum_{i=1}^{k} 1/v_i}.$$

Since these methods are already implemented in two common meta-analysis computting packages, it is not necessary for the research reviewer to understand these details.

Restricted maximum likelihood and full maximum likelihood estimates require an iterative solution. These can be programmed in R; an example code is provided in the Appendix. Estimates of the variance component can also be obtained using HLM software (Raudenbush et al. 2004), following the directions for estimating a v-known model. (Note that in the latest version of HLM, the program needs to be run from a DOS prompt in batch mode). These estimates can also be obtained using SAS Proc Mixed; a sample program is included in the Appendix. The program R can also be used conducting a simple iterative analysis to compute the restricted maximum likelihood estimate of the variance component. The Appendix contains a program that computes the overall variance component as given in Table 3.1. Note that the most common method for computing the variance component remains the method of moments. Most research reviewers do not have the computing packages needed to obtain the REML estimate.

Once the reviewer has made the decision about fixed or random effects models, and then computed the variance component, if necessary, the analysis proceeds by first computing the weighted mean effect size (fixed or random), and then testing for homogeneity of effect sizes. In the fixed effects case, this requires the computation of the Q statistic as outlined in Chap. 2. In the random effects case, this requires the test that the estimated variance component, τ^2, is different from zero as seen in Chap. 2. More detail about these analyses can be found in introductory texts (Borenstein et al. 2009; Higgins and Green 2011; Lipsey and Wilson 2000).

3.4.1 Example

Sirin (2005) reports on a meta-analysis of studies estimating the relationship between measures of socio-economic status and achievement. The socio-economic status measures used in the studies in the meta-analysis include parental income, parental education level, and eligibility for free or reduced lunch. The achievement measures include grade point average, achievement tests developed within each study, state developed tests, and also standardized tests such as the California Achievement Test. Below is an illustration of the different options for conducting a random effects and mixed effects analysis with categorical data. There are eleven studies in the subset of the Sirin data that use free lunch eligibility as the measure of SES. Five of these studies use a state-developed test as a measure of achievement, and six use one of the widely used standardized achievement tests such as the Stanford or the WAIS. Table 3.1 gives the variance components as computed by the DerSimonian and Laird (1986) method (also called the method of moments), and SAS (the programs are given in the Appendix to this chapter). The SAS estimates use restricted maximum likelihood. Note that I provide both the common estimate of the variance component, and separate estimates within the two groups.

Table 3.1 Comparison of two methods to compute random effects variance

Test type	n	Q	Method of moments estimate	SAS estimate using REML
State test	5	60.99[a]	0.0872	0.086
Standardized achievement test	6	75.14[a]	0.0247	0.0417
Total	11	514.86[a]	0.111	0.0957

[a] $p < 0.05$, indicating significant heterogeneity

Table 3.2 Analysis with a single variance component estimate

Type of achievement test	n	Weighted mean	SD of mean effect size	Lower 95% CI	Upper 95% CI
State	5	0.614	0.146	0.327	0.900
Standardized	6	0.265	0.129	0.012	0.517
Total	11	0.417	0.096	0.228	0.606

In this example, both groups have significant variability among the effect sizes, indicating that the variance components are all significantly different from zero. We also see that the method of moments estimate and the estimate using REML differ in the case of the standardized achievement tests, complicating both the choice of estimation method and of whether to use separate variance component estimates within each group or a common estimate.

As Borenstein et al. (2009) point out, the estimates for the variance component are biased when samples are small. Given the potential bias in the estimates and the differences in the two estimates for the studies using standardized achievement tests, the analysis in Table 3.2 uses the SAS estimate of the common variance component, $\tau^2 = 0.0957$ to compute the random effects ANOVA. While the state tests have a mean Fisher's z-score that is larger than that for the standardized tests, their confidence intervals do overlap indicating that these two means are not significantly different.

3.5   Confounding of Moderators in Effect Size Models

As mentioned earlier, there are several examples of meta-analyses that do not follow standard statistical practice. One example concerns the use of multiple statistical tests without adjusting the Type I error rate. Many of the meta-analyses in the published literature report on a series of one-way ANOVA models when examining the effects of moderators (Ehri et al. 2001; Sirin 2005). Often the results of examining one moderator at a time are provided, with confidence intervals for each of the mean effect sizes within groups, and the results of an omnibus test of significance among the mean values. Examining a series of one-way ANOVA models in meta-analysis has all the same difficulties as conducting these in any other statistical analysis context. Primary studies rarely report one-way ANOVAs, relying more on multivariate analyses such as multi-factor ANOVA or regression. Why the meta-analytic literature has not followed these recommendations is not clear.

There are a number of reasons why meta-analysts need to be careful about reporting a series of single-variable analyses. The first is the issue of confounding moderators. It could easily be the case that the mean effect sizes using different measures of a construct are significantly different from one another, and that there are also significant differences among the means for groups of studies whose participants are of different age ranges. If we only conduct these one-way analyses without examining the relationship between age of participants and type of measure, we will not know if these two variables are confounded. Related to this problem is one of interpretation. How do we translate a series of one-way ANOVAs results into recommendations for practice? In our example, which of the two moderators are more important? Should we recommend that only one type of measure be used? Or, should we focus on the effectiveness of the intervention for particular age groups?

A final issue relates to the problem of multiple comparisons. As we learned in our first statistics course, conducting a series of statistical tests all at the $p = 0.05$ level will increase our chances of finding a spurious result. The more statistical tests we conduct, the more likely we will find a statistically significant result by chance. But, we do not seem to heed this advice in the practice of meta-analysis. We often see a series of analyses reported, each testing the significance of the mean effect size or the between-group homogeneity test. Fortunately, more recent research syntheses have also reported on the confidence intervals for these means, obviating the problems that may occur with singular reliance on statistical tests. Hedges and Olkin (1985) discuss the adjustment of the significance level for multiple comparisons using Bonferroni methods, but few meta-analyses use these methods across the meta-analysis itself.

What should meta-analysts do when trying to examine the relationship of a number of moderators to effect size magnitude? The first is to recognize that moderators are bound to suffer from confounding given the nature of meta-analysis. Especially in the social sciences, the studies in the synthesis are rarely replications of one another, and use various samples, measures and procedures. Research reviewers should examine the relationships among moderators. These relationships can be examined using correlations, two-way tables of frequencies, or other methods. Understanding the patterns of moderators across studies will not only help researchers and readers understand how to interpret the moderator analyses, it will also highlight the nature of the literature itself. It could be that no study uses the highest quality measure of a construct with a particular sample of participants, and thus, we do not know how effective an intervention is with that sample.

Research synthesists should also focus more on the confidence interval for the mean effect sizes within a given grouping of studies, rather than on the significance tests. The overlap among the confidence intervals for the mean effect sizes will provide the same information as the statistical significance test, but is not subject to the problems with multiple comparisons (Valentine et al. 2010).

More researchers should also use meta-regression to examine the relationship of multiple predictors on effect magnitude. Once a researcher has explored the possible confounding among moderators in the literature, a set of moderators could be used with meta-regression to see how they relate to effect size net of the other variables in the effect size model. Lipsey (2009) advocates the use of hierarchical regression; the first set of predictors in the model may be control variables such as type of measure used, and then the moderators of substantive interest are entered to examine how much of the residual variation is accounted for by these predictors. If one of the goals for research synthesis is to provide evidence for practice and policy, we need to understand what contextual and study level moderators may explain the differences among studies in their results.

Table 3.3 One-way fixed effects ANOVA models based on Sirin (2005)

Moderator	n	z (sd)	Lower CI	Upper CI	Q_B	p
Achievement						
GPA	10	0.26(0.011)	0.24	0.29	$Q_B = 23.73$	<0.001
Achievement	5	0.27(0.019)	0.24	0.31		
State	6	0.21(0.005)	0.20	0.22		
Standardized	12	0.22(0.012)	0.20	0.24		
SES						
Free lunch	11	0.24(0.014)	0.21	0.26	$Q_B = 27.24$	<0.001
Income	4	0.21(0.005)	0.20	0.22		
Education	18	0.26(0.009)	0.25	0.28		
Total	33	0.22(0.004)	0.22	0.23		

Table 3.4 Crosstabulation of SES and Achievement

Achievement measures	SES measures		
	Free lunch	Income	Education
GPA	1	0	9
	9.1%	0%	50%
Achievement	0	1	4
	0%	25%	22.2%
State	4	2	0
	36.4%	50%	0%
Standardized	6	1	5
	54.5%	25%	27.8%
Total	11	4	18
	100%	100%	100%

3.5.1   Example

In the Sirin (2005) study, the results of a number of one-way ANOVA models are reported. For example, in Table 3.3, I present a subset of the Sirin studies, examining the mean effect size for groups defined by type of achievement measure and by type of SES measure. These studies are only a subset of those from the paper, and do not necessarily reflect the results of the original analysis; they are here for illustration purposes only. Since the between-group Q_B test is significant for both achievement and SES measures, we can state that there is at least one effect size within each grouping that is not equal to the other means.

What is not clear from this table is whether these two variables, type of achievement measure and type of SES measure are associated with one another. Table 3.4 below shows the crosstabulation for the number of effect sizes within the cells defined by these two variables. The Pearson's chi-square test of independence is significant, $\chi^2(6) = 16.67$, p = 0.011. The chi-square test indicates a relationship between type of SES measure and type of achievement measure. Examining the table, we see that

Table 3.5 Random effects mean effect sizes, and 95% confidence intervals for studies classified by Achievement and SES measures

	SES measures		
Achievement measures	Free lunch	Income	Education
GPA	0.41	–	0.16
	[0.02, 0.81]		[0.03, 0.30]
	k = 1		k = 9
Achievement	–	0.46	0.28
		[0.01, 0.91]	[0.07, 0.49]
		k = 1	k = 4
State	0.54	0.20	–
	[0.32, 0.76]	[−0.07, 0.47]	
	k = 4	k = 2	
Standardized	0.26	0.15	0.30
	[0.10, 0.43]	[−0.24, 0.54]	[0.13, 0.48]
	k = 6	k = 1	k = 5

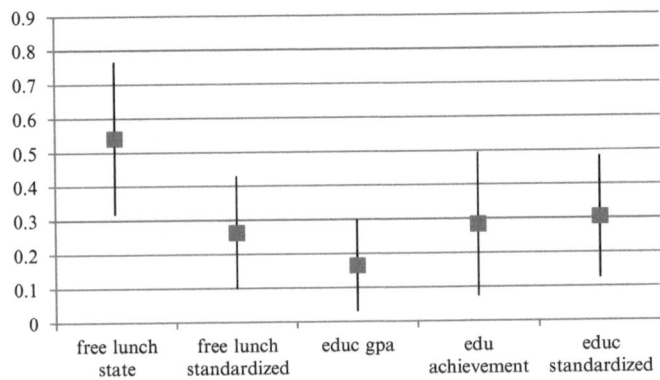

Fig. 3.2 Error bar plot for random effects means in Sirin (2005)

most of the studies using GPA for achievement also use education as the proxy for SES. Studies employing state tests tended to use free lunch status as the proxy for SES, with the remainder of those studies using income.

The results in Table 3.5 show the random effect size means for the groups defined by both these variables simultaneously. What we do see is that the four studies using both state tests and free lunch as the SES proxy have a weighted mean effect size that is almost twice as large as the other cells that have at least two effects. Figure 3.2 provides an error bar chart for the means and their 95% confidence interval for the studies using either free lunch or education as the measure of SES. Though all the confidence intervals overlap, there is some evidence that the studies using free lunch for SES and the state test for achievement have a larger estimate of the correlation between SES and achievement. For this subset of studies, there is an interaction between two moderator variables that can influence the interpretation of results.

This small example illustrates the need for reviewers to examine carefully the relationships among moderators. As indicated above, meta-regression is a recommended strategy when there are sufficient numbers of effect sizes since we can examine the influence of multiple moderators on effect size at the same time. If meta-regression is not possible, then the reviewer needs to provide evidence that confounding among the target moderators will not impact the interpretation of the findings.

3.6 Conducting a Meta-Regression

When a reviewer has a larger sample of studies, conducting a meta-regression can help a reviewer avoid conducting multiple significance tests, and instead examine the conditional relationship among the predictors and effect size magnitude. Introductory text books (Borenstein et al. 2009; Lipsey and Wilson 2000) provide more detail about the background for conducting a meta-regression; here I will provide an example of meta-regression using another subset of the Sirin (2005) data. The Appendix provides both SPSS and SAS sample programs for computing the model discussed below.

We conduct a meta-regression when we want to examine how a set of p predictor variables relate to the variation among effect sizes. Our hypothetical linear model can be written as

$$\theta_i = \beta_0 + \beta_1 x_{i1} + \ldots + \beta_p x_{ip},$$

where θ_i is the effect size in study i, the $x_{i1},\ldots,x_{ip}$ are the values of the predictor variable for study i, and the $\beta_0,\ldots,\beta_p$ are the unknown regression coefficients. In the fixed effects case, we will use our values of T_i as our estimates of the θ_i with v_i as the variance of these effect size estimates. We use weighted least squares regression to estimate the model where our weights are equal to $w_i = 1/v_i$. For a random effects meta-regression, the weights are $w_i^* = 1/(v_i + \tau^2)$ where τ^2 is estimated from one of the methods from Sect. 3.4 (For the interested reader, Chap. 8 illustrates how to estimate the random effects meta-regression using restricted maximum likelihood methods).

3.6.1 Example

Below I provide sample output from SPSS to illustrate how fixed effects meta-regression analysis is conducted (The Appendix contains the steps needed to conduct the meta-regression in SPSS, as well as a sample program for SAS).

Model Summary

Model	R	R Square	Adjusted R Square	Std. Error of the Estimate
1	.712[a]	.506	.450	5.48726

a. Predictors: (Constant), educ, percmin, freelunch, grade

ANOVA[b,c]

Model		Sum of Squares	df	Mean Square	F	Sig.
1	Regression	1081.426	4	270.356	8.979	.000[a]
	Residual	1053.852	35	30.110		
	Total	2135.278	39			

a. Predictors: (Constant), educ, percmin, freelunch, grade
b. Dependent Variable: ztrans
c. Weighted Least Squares Regression - Weighted by wt

Coefficients[a,b]

Model		Unstandardized Coefficients		Standardized Coefficients	t	Sig.
		B	Std. Error	Beta		
1	(Constant)	.041	.081		.504	.618
	percmin	-.001	.001	-.224	-1.879	.069
	grade	.073	.024	.377	2.983	.005
	freelunch	.354	.068	.649	5.193	.000
	educ	.104	.059	.223	1.752	.088

a. Dependent Variable: ztrans
b. Weighted Least Squares Regression - Weighted by wt

Fig. 3.3 SPSS output for Sirin (2005) meta-regression example

The data used in this example are given in the Data Appendix at the end of the book. Let us say that we are interested in examining how the values of Fisher's z-transformation in the Sirin data are related to the grade level of the students in the study, the percent of minority students represented in the study's sample (percmin), and the way that income was measured in the study. For this example, I have included two dummy codes, free lunch, which indicates that the study used eligibility for free lunch to identify low-income students, and education, which indicates that the study used the level of education of the parent as a indicator of income. The default SPSS output is given in Fig. 3.3.

The first table in Fig. 3.3 provides the Model Summary. This table should be ignored. The adjusted R-square value cannot be interpreted as in usual practice since, as Konstantopolous and Hedges (2009) point out, the maximum value for the population multiple correlation in this case is often much less than one. The second table provides the summary of the F-tests for the meta-regression. In this table,

Table 3.6 Corrected table for meta-regression results using Sirin (2005)

Coefficient	Estimate	SE_j	S_j	Z_j	p-value
Intercept	0.041	0.081	0.0150	2.73	0.003
% minority	−0.001	0.001	0.0002	−5.00	0.000
Grade	0.073	0.024	0.0043	16.98	0.000
Free lunch	0.354	0.068	0.0120	29.50	0.000
Education level	0.104	0.059	0.0110	9.45	0.000

the Sum of Squares for the Residual is the statistic called Q_E, which provides a test of the goodness of fit of the linear model. This statistic is the weighted residual sum of squares about the regression line. When the statistic is significant, the variation around the regression line is greater than expected given that the model is a good fit to the data. The statistic Q_E is compared to a chi-square distribution with $(k - p - 1)$ degrees of freedom. In this case, we have $Q_E = 1{,}053.85$, compared to the 95% critical value of a chi-square with 34 degrees of freedom which is equal to 48.60. Thus, we have greater than expected residual variance around our regression line, indicating that our model is not a good fit to the data.

The third table presents the estimated regression coefficients. As many textbooks point out (Borenstein et al. 2009; Konstantopolous and Hedges 2009; Lipsey and Wilson 2000), the standard errors and tests of significance for the coefficients printed in many weighted regression routines in common statistical packages are incorrect since they are based on a model slightly different than the one used for fixed-effects meta-regression. The corrected standard errors and tests of significance must be computed by hand. The corrected standard errors are given as

$$S_j = \frac{SE_j}{\sqrt{MSE}} \tag{3.1}$$

where the SE_j are the printed standard errors, and MSE is the mean square error given in ANOVA table for the regression. The correct tests of significance use the Z distribution. If we wish to test that a given regression coefficient is equal to 0, we use a two-sided test where we compute the statistics Z_j given by

$$Z_j = \frac{|b_j|}{S_j}$$

where b_j is the estimated regression coefficient, and S_j is the corrected standard error given in (3.1). Table 3.6 provides the corrected standard errors and tests of significance for the Sirin data.

In this example, where $k = 39$, all of the coefficients are significantly different from zero. We see that the larger the percent minority included in the sample, the smaller the Fisher z-transformation (and by extension, correlation) between socio-economic status and achievement. Higher grade levels have larger Fisher z's as do studies where free lunch or education level are used as the indicators of income. The comparison category to the indicator of income are rating scales of the occupational status and actual reported income. Though the value of the Q_E statistic indicates that significant residual variation remains, this model does provide insight into the variation among the effect sizes in this data.

3.7 Interpretation of Moderator Analyses

As seen in the example above, reviewers need to exercise caution in their interpretation of the results of moderator analyses when potential confounding occurs. Discovering what variables may be confounded is difficult in meta-analysis since many reviews include only a small number of studies. However, the example provided above adds to our understanding of the complex relationships among measures of SES and source of the achievement measures.

Adding to the complexity of the interpretation of moderator analyses is the nature of these analyses themselves. Research reviewers do not have control over the population of studies and how these studies were conducted. Thus, unlike a primary study, we cannot randomly assign particular studies to using just males or just females, or to assessing achievement only using a single measure. The analyses in a research synthesis are exploratory and observational. We usually have little evidence from a meta-analysis about why study characteristics relate to effect magnitude. We can only describe what the relationships are among achievement and SES measures, for example. These analyses are observational only, and do not provide information about why or how these relationships operate. Chapter 9 discusses the nature of the inferences that are possible from a meta-analysis.

Preliminary stages of meta-analysis should conform to standard statistical analysis practice, such as providing a plot of the data, choosing a priori the moderators that are of most interest, and exploring homogeneity of the effect sizes. Choosing these analyses ahead of time ensures that the research reviewer does not capitalize on chance by conducting a series of analyses without adjusting the overall Type I error rate. In addition, it is difficult to interpret a series of single variable analyses since we do not know which moderator should be considered important. As in the example above, we find that studies using state tests tend to use eligibility for free lunch as the measure of SES, a measure that is not as sensitive as income. The ultimate goal of a meta-analysis is to help clarify what is known about a given intervention or phenomenon, and the analyses we present should aim at illuminating what is and is not yet known in a given research area. Carefully interpreting the analyses is ultimately the responsibility of the reviewer.

Appendix

Computing the Variance Component Using SAS

The example in Sect. 3.4.1 involves the computation of the variance component using restricted maximum likelihood. Below is a SAS program that computes the estimates for the analyses in Tables 3.1–3.3. The first program enters the data used in this example. The variables are the study name, the correlation, Fisher's z corresponding to the correlation, the variance of Fisher's z, a code that takes the value 1 for studies using state tests for achievement, and 2 for standardized tests, and the study weight. The second program provides the code to compute the variance component using restricted maximum likelihood. The first line calls Proc Mixed, with **cl** proving the confidence limits, and **method** indicating the used of restricted maximum likelihood. The **model** statement indicates that Fisher's z is the outcome, and the model is a simple random effects model with no predictors. The options **S** and **CL** provide the fixed-effects parameter estimates, and the confidence interval, respectively. The **class** statement indicates that study is a class variable, which will be designated as the random effect later in the code. The **random** statement designates the random effect, and the option **solution** prints the estimate of the random effect, or in this case, the random effects variance. The **repeated** statement specifies the covariance structure of the error term. For meta-analysis, we use the option **group** to allow between-group (in this case, study) heterogeneity. The **parms** statement provides the starting values for the covariance parameters. The first value is the starting value for the overall variance component for this model. The next 11 elements are the within-study estimates of the variance of the effect size. The option **eqcons** fixes the variances for the 11 studies in the analysis since these are considered known in a meta-analysis model.

```
data sirinch6ex;
input study $ corr zval varz achmeas wt;
cards;
```

alspurb	.7190	.9055717	.0278	1	36.00
alsprur	.0720	.0721248	.0097	1	103.00
calban1	.6800	.8291140	.0008	1	1298.00
dixflo	.4670	.5062267	.0122	1	82.00
grelan	.6500	.7752987	.0086	1	116.00
gulbur	.1240	.1246415	.0006	2	1570.00
johnlin	.1750	.1768200	.0006	2	1683.00
kling	.5400	.6041556	.0030	2	329.00
recste	.0600	.0600722	.0077	2	130.00
schu	.4300	.4598967	.0077	2	130.00
shawal	.1660	.1675505	.0030	2	332.00

```
;
run;
```

```
proc mixed cl method=reml data=sirinch6ex;
class study;
model zval =/S CL;
random int/SUBJECT=study S;
repeated/GROUP=study;
Parms (0.01 to 2.00 by 0.01)
(.0278) (.0097) (.0008) (.0122)
(.0086) (.0006) (.0006) (.0030)
(.0077) (.0077) (.0030)
/EQCONS=2 to 12;
run;
```

Computing the Variance Component Using R

Below is a simple iterative program to obtain the restricted maximum likelihood estimates from R. The estimates are described by Raudenbush (2009). The first two lines set up the starting values for the random effects mean and variance, namely a vector of 100 elements, all having the value of zero. The 100 iterations take place within the brackets, "{ }". First, the value for the new variance for the vector of effect sizes is computed as adding our current value of the random effects variance, sigrmle[i] to the within-study variation, sirin$ztrans. Given this value of the vector of the within-study variances, vstar, we recomputed a new value of the random effects mean, beta0. Given this value of the random effects mean, beta0, we recalculate the value of the variance component sigrmle. We cycle back and forth for 100 iterations in this program. Below the program are the results from R using this example.

```
beta0<−rep(0, c(100))
sigrmle<−rep(0,c(100))
for (i in 1:100) {
vstar<−sirin$var+sigrmle[i]
beta0[i]<−sum(sirin$ztrans/vstar)/sum(1/vstar)
sigrmle[i+1]<−(sum(((((sirin$ztrans-beta0[i])**2)-sirin$var)/(vstar**2))/sum(1/
(vstar**2)))+(1/sum(1/vstar))
}
> beta0
[1] 0.3551265 0.4162903 0.4169039 0.4169685 0.4169757 0.4169765 0.4169766
[8] 0.4169766 0.4169766 0.4169766 0.4169766 0.4169766 0.4169766 0.4169766

. . . . . . . . ...
> sigrmle
[1] 0.00000000 0.08408039 0.09172159 0.09258992 0.09268741 0.09269834
[7]      0.09269957     0.09269971     0.09269972     0.09269972     0.09269972
0.09269972. . . . . ..
```

We can see that by the 8th iteration, both the variance component and the random effects mean do not change. The estimates from R correspond to those obtained using SAS.

Computing the Fixed Effects Meta-regression Using SPSS

The analysis in Table 3.6 can be computed from the Analyze menu in SPSS, under Regression, and then under Linear Regression. The Dependent variable will be the Fisher z-transformations of the correlation coefficients. In this example, the Independent variables are the percent minority in the sample, the dummy code indicating whether free lunch was used as the measure of socio-economic status, the dummy code indicating whether parent education level was used as the measure of socio-economic status, and grade level. The WLS Weight for the analysis is the inverse of the Fisher's z-transformation variance, or $1/(n-3)$.

Computing the Fixed Effects Meta-regression Using SAS

To compute the fixed effects meta-regression in SAS, we use Proc Reg as illustrated in the code given below. The **model** statement begins with the outcome, the Fisher z-transformation, an equals sign, and the list of predictors. An option to the **model** statement is given by the slash /and followed by **I** which indicates that we want the inverse of the crossproducts matrix. The inverse of the crossproducts matrix will provide the correct variances for the regression coefficients. The **weight** statement indicates that the variable wt should be used in the weighted least squares regression.

```
proc reg data=Chap. 3;
model ztrans=grade percmin freelunch educ/I;
weight wt;
print;
run;
```

The output provided by SAS is first the inverse of the crossproducts matrix:

X'X Inverse, Parameter Estimates, and SSE

	Intercept	Grade	% minority	Free lunch	Parent's education
Intercept	0.000218	0.0000594	−4.746E-7	−0.0000666	−0.0000709
Grade	−0.0000594	0.0000198	3.834E-8	0.0000126	0.0000143
% minority	−4.745E-7	3.834E-8	9.403E-9	7.456E-8	5.824E-8
Free lunch	−0.0000666	0.0000126	7.456E-8	0.000154	0.0000362
Parent's education	−0.0000709	0.0000143	5.824E-8	0.0000362	0.000117

The diagonal elements of this matrix are the correct variances for the weighted regression coefficients. For example, the correct standard error for Parent's education level is $\sqrt{0.000117} = 0.0108$, which corresponds to that computed using the adjusted SPSS values in Table 3.6. Next, SAS provides the ANOVA table for the regression and the R-squared values.

Analysis of Variance

Source	DF	Sum of squares	Mean square	F value	Pr > F
Model	4	1081.426	270.356	8.98	<.0001
Error	35	1053.852	30.110		
Corrected total	39	2135.278			
Root MSE	5.487	R-Square	0.5065		
Dependent mean	0.272	Adj R-Sq	0.4501		
Coeff var	2014.675				

As in the example with SPSS, we will use the square root of the mean square error given in the output above to correct the standard errors and the significance tests of the regression coefficients given below.

Variable	DF	Parameter estimate	Standard error	t-value	Pr >
Intercept	1	0.0408	0.0811	0.50	0.6176
Grade	1	0.0729	0.0244	2.98	0.0052
% minority	1	−0.000999	0.00053	−1.88	0.0686
Free lunch	1	0.353	0.0681	5.19	<.0001
Parent's education	1	0.104	0.0594	1.75	0.0885

References

Anderson, L.M., M. Petticrew, E. Rehfuess, R. Armstrong, E. Ueffing, P. Baker, D. Francis, and P. Tugwell. 2011. Using logic models to capture complexity in systematic reviews. *Research Synthesis Methods* 2: 33–42.

Barel, E., M.H. Van IJzendoorn, A. Sagi-Schwartz, and M.J. Bakermans-Kranenburg. 2010. Surviving the Holocaust: A meta-analysis of the long-term sequelae of a genocide. *Psychological Bulletin* 136(5): 677–698.

Borenstein, M., L.V. Hedges, J.P.T. Higgins, and H.R. Rothstein. 2009. *Introduction to meta-analysis*. Chicester: Wiley.

Cochrane Information Management System. 2011. RevMan 5.1. Oxford, UK: Cochrane Collaboration.

Comprehensive Meta-Analysis Version 2. 2006. Englewood, NJ: Biostat.

Cooper, H., L.V. Hedges, and J.C. Valentine (eds.). 2009. *The handbook of research synthesis and meta-analysis*. New York: Russell Sage.

Dersimonian, R., and N. Laird. 1986. Meta-analysis in clinical trials. *Controlled Clinical Trials* 7: 177–188.

Ehri, L.C., S. Nunes, S. Stahl, and D. Willows. 2001. Systematic phonics instruction helps students learn to read: Evidence from the National Reading Panel's meta-analysis. *Review of Educational Research* 71: 393–448.

Greenland, S., J. Pearl, and J.M. Robins. 1999. Causal diagrams for epidemiologic research. *Epidemiology* 10(1): 37–48.

Hedges, L.V., and I. Olkin. 1985. *Statistical methods for meta-analysis*. New York: Academic.

Hedges, L.V., and J.L. Vevea. 1998. Fixed- and random-effects models in meta-analysis. *Psychological Methods* 3(4): 486–504.

Higgins, J.P.T., and S. Green. 2011. *Cochrane handbook for systematic reviews of interventions*. Oxford, UK: The Cochrane Collaboration.

Joffe, M., and J. Mindell. 2006. Complex causal process diagrams for analyzing the health impacts of policy intervention. *American Journal of Public Health* 96(3): 473–479.

Konstantopolous, S., and L.V. Hedges. 2009. Analyzing effect sizes: Fixed-effects models. In *The handbook of research synthesis and meta-analysis*, 2nd ed, ed. H. Cooper, L.V. Hedges, and J.C. Valentine. New York: Russell Sage.

Lipsey, M.W. 2009. Identifying interesting variables and analysis opportunities. In *The handbook of research synthesis and meta-analysis*, 2nd ed, ed. H. Cooper, L.V. Hedges, and J.C. Valentine. New York: Russell Sage.

Lipsey, M.W., and D.B. Wilson. 2000. *Practical meta-analysis*. Thousand Oaks: Sage Publications.

Littell, J.H., M. Campbell, S. Green, and B. Toews. 2005. Multisystemic therapy for social, emotional and behavioral problems in youth aged 10–17. *Cochrane Database of Systematc Reviews* (4). doi:10.1002/14651858.CD004797.pub4.

Raudenbush, S.W. 1983. Utilizing controversy as a source of hypotheses for meta-analysis: The case of teacher expectancy's effects on pupil IQ. In *Evaluation studies review annual*, 8th ed, ed. R.J. Light. Beverly Hills: Sage Publications.

Raudenbush, S.W. 2009. Analyzing effect sizes: Random-effects models. In *The handbook of research synthesis and meta-analysis*, 2nd ed, ed. H. Cooper, L.V. Hedges, and J.C. Valentine. New York: Russell Sage.

Raudenbush, S.W., A.S. Bryk, Y.F. Cheong, and R. Congdon. 2004. *HLM 6 for Windows*. Lincolnwood: Scientific Software International.

Rosenthal, R., and L. Jacobson. 1968. *Pygmalion in the classroom*. New York: Holt, Rinehart & Winston.

Sirin, S.R. 2005. Socioeconomic status and academic achievement: A meta-analytic review of research. *Review of Educational Research* 75(3): 417–453. doi:10.3102/00346543075003417.

Valentine, J.C., T.D. Pigott, and H.R. Rothstein. 2010. How many studies do you need? A primer on statistical power in meta-analysis. *Journal of Educational and Behavioral Statistics* 35: 215–47.

Chapter 4
Power Analysis for the Mean Effect Size

Abstract This chapter provides methods for computing the a priori power of the test of the mean effect size. Both fixed and random effects models tests are discussed. In addition, examples are provided for computing the number of studies needed to detect a substantively important effect size, and the detectable effect size with a given number of studies.

4.1 Background

Researchers planning new primary studies use power analysis to increase their odds of finding results that they believe are present. Power analysis allows the researcher to determine a priori the number of participants needed in a study to find statistical significance. For example, if we want to know that an intervention has a substantive effect, we need to plan a study with sufficient numbers of participants, and adequately sensitive measures and analysis strategies to detect that effect. Few primary researchers would risk conducting an intervention study without a power analysis.

In research synthesis and meta-analysis, reviewers plan systematic reviews in areas that have matured enough to have a sufficient body of evidence, and where a careful review may enhance the overall understanding of a field. Unlike primary researchers, reviewers cannot plan in advance the sample size (in this case, the number of primary studies) needed to find important substantive effects. Thus, on the surface, power analysis in meta-analysis, especially a priori power analyses, might appear a futile exercise.

I take an alternative view of the value of a power analysis in research synthesis. As emphasized in Chap. 3, knowledge of an area is critical to the planning of a synthesis, especially with regard to planning the moderator analyses carefully so as to avoid Type I errors. The same could be argued for the importance of a power analysis as preliminary to a systematic review. Though a reviewer might not know the exact number of studies that will ultimately be relevant to the review, the reviewer should have a clear understanding of the size of the effect that would be

T.D. Pigott, *Advances in Meta-Analysis*, Statistics for Social and Behavioral Sciences, DOI 10.1007/978-1-4614-2278-5_4, © Springer Science+Business Media, LLC 2012

considered substantively important and of the statistical methods (such as moderator analyses) that will be used. Examining the potential power of a meta-analysis prior to collecting and evaluating the studies will add strength to the reviewer's later decisions about conducting particular analyses. Say, for example, we are interested in understanding the potential magnitude of an adverse effect of a given medical treatment, but believe the incidence of this effect is rare. The reviewer would know a priori that an estimate of this adverse effect cannot be examined if there are not a sufficient number of studies. In an educational setting, if we are interested in understanding the magnitude of the effect of a dropout prevention program, we might find that we do not have enough studies to detect, for example, a standardized mean difference of 0.5 between the treatment and control groups. This information can then provide a rationale for the statistical analyses conducted (or not) with the effect size data.

In order to conduct a power analysis a priori, the reviewer must be transparent about the assumptions he or she is making at the outset of the synthesis. This transparency can serve as a check for the analysis of the effect size data so that reviewers do not conduct too many statistical analyses that may capitalize on chance. Reviewers can also inform policymakers prior to a synthesis of the numbers of studies that are needed to find a significant effect of an intervention. There are circumstances where we do not have enough primary research to make an informed, research-based decision, and an a priori power analysis of a systematic review would provide empirical evidence for that claim.

As mentioned above, the reviewer conducting a power analysis in meta-analysis needs to have informed judgments about important characteristics of studies, such as the typical within-study sample size, and the magnitude of a substantively important effect size. Since reviewers need to make assumptions about these quantities, there is a temptation to compute power after analyzing the data. The use of retrospective power analyses is not new; much has been written in the medical, natural and behavioral sciences about the limitations and advantages of power analysis after the study is complete (Goodman and Berlin 1994; Hayes and Steidl 1997; Reed and Blaustein 1995; Thomas 1997; Zumbo and Hubley 1998). In many fields, retrospective power analysis has been suggested as a way to interpret null results. For example, one often-cited study of amphibian populations by Reed and Blaustein (1995) concludes that prior studies finding a lack of decline in populations "cannot be supported statistically" (p. 1299) since the statistical power in these studies ranges from less than 0.06 to 0.45. In medicine, Freiman et al. (1978) computed the retrospective power of 71 "negative" trials, finding that a majority of studies had power levels less than 0.80 for detecting a 25% or a 50% reduction in mortality.

Critics of retrospective power analysis (Goodman and Berlin 1994; Hayes and Steidl 1997; Zumbo and Hubley 1998) point out the logical flaws in using the observed results to compute retrospective power. The major issue is that power is related to statistical significance; once a researcher obtains a finding of no significant difference, then the researcher knows that power was inadequate. The observed power, power calculated after the analysis is completed, does not provide any more evidence than the p-value for a given test. Power makes sense only in relation to a specific effect size; in other words, power is computed as the power to detect a specific difference between groups, or size of association between constructs. It is redundant to compute power for a non-significant test since we already know that our test is underpowered for our obtained result. For researchers using research synthesis, power computations need to occur in the early stages of the review to determine if adequate evidence exists to test a given hypothesis. As I will discuss later, reviewers should start with substantively significant values for an average effect size and compute power given a range of values for the effect size variances, sample sizes within studies, and number of studies.

The following chapters provide an overview of power analysis in meta-analysis, and illustrate its use with examples. This chapter begins with a review of power analysis in general and then in meta-analysis, focusing on tests of the mean effect size, under both the fixed and random effects model.

4.2 Fundamentals of Power Analysis

All power analyses are conducted with a specific statistical test in mind, and are described in relation to a null hypothesis and a particular alternative hypothesis. Though organizations such as the American Psychological Association (American Psychological Association 2009) are emphasizing the interpretation of effect sizes rather than statistical significance, power computations remain in the realm of the null and alternative hypothesis. In order to compute power, researchers use a value for the alternative hypothesis that represents some substantively important quantity, such as the smallest effect size that would be considered substantively critical, and then discuss the power of a statistical test under the specific alternative hypothesis posed.

The computation of power analysis for most statistical tests depends on four general quantities: the sample size of the study, the substantively important effect size that will be tested against the null hypothesis, the desired significance level of the statistical test, and the desired level of power (Cohen 1988; Murphy and Myors 2004). Any three of these four quantities can be used to solve for the fourth one. For example, researchers often ask about the number of participants needed to find a given difference between the treatment and control groups. With the particular value for the difference between the groups, the desired significance level and power level, the researcher can find the optimal sample size. Alternatively, the researcher may know how many participants are available, and can compute the power level for the test given a particular value for the group difference.

When we compute power, we are actually comparing two distributions: that of our test statistic given that the null hypothesis is true, and that of our test statistic when the alternative hypothesis is correct. The power of our test is the probability that our obtained test statistic will be considered significant under the null

Table 4.1 Steps for determining power in meta-analysis

1. Establish a critical value for statistical significance, c_α
2. Posit an overall effect size, θ
3. Estimate the number of studies included in the meta-analysis, k
4. Decide on fixed or random effects models
   a. Posit "typical" within-study samples sizes and effect size variances for fixed effects
   b. For random effects also posit typical variance component, τ^2
5. Compute power for specific tests in a meta-analysis

Table 4.2 Information needed to compute power in meta-analysis

Information needed at the level of the research synthesis

- Critical value/criterion for statistical significance, c_α
- Effect size of practical significance, θ
- Number of studies for the meta-analysis, k
- For random effects: variance component (between-studies variance), τ^2

Information needed at the level of the studies included in the research synthesis

- Within-study sample sizes, N
- Study effect size variances (related to sample size), v

hypothesis when the alternative hypothesis is really true. Thus, power depends on both the null and the alternative hypothesis in a given context, and is typically decided upon prior to computing the test.

Computation of power in a meta-analysis is similar to that for other statistical tests. For example, we need the number of studies in the review (the "sample size"), the anticipated overall mean effect size, and the criterion for statistical significance. Since our unit of analysis is a primary study, power in meta-analysis also depends on the sample size within studies. Table 4.1 outlines the steps to compute the power of a meta-analysis. The first three steps are analogous to those for power computations in primary studies; the fourth step requires knowledge about the size of the sample and the effect sizes of the individual studies. These quantities are usually variable across studies, and requires the reviewer to have deep knowledge about the types of studies typically conducted in a given research area in order to make accurate assumptions for computing power. Table 4.2 provides more details about the information needed for computing power at two levels of a systematic review: at the level of the review itself, and at the level of the individual studies.

The rest of this chapter covers power for the statistical test of the mean effect size, in both fixed and random effects models. The power for the mean effect size will be important for planning reviews when the interest is in the overall effect of an intervention, an estimate of the mean association between two constructs, or the overall incidence rate of a condition. For example, a policymaker may be interested in knowing if the average effect of an expensive intervention will be worth the cost. A reviewer might then want to know the minimum number of studies needed to detect a substantively important effect size that would justify the cost of the intervention. Alternatively, I may want to know the size of the effect I could detect with a given number of studies – and whether this detectable effect will be of substantive interest.

4.3   Test of the Mean Effect Size in the Fixed Effects Model

The first step in any meta-analysis is to examine the mean effect size. In planning a research synthesis, reviewers may be interested in understanding the average effect for an intervention, or the average incidence rate across studies. In any given area of research, a reviewer should have a good understanding of the size of an effect that would be considered substantively important. The reviewer can then compute the minimum number of studies needed to detect that effect size with a particular power. Alternatively, the reviewer might have a rough idea of the number of studies that exist, perhaps because of the results of a prior review. This information could help a reviewer to compute the size of an effect that will be able to be detected with a particular power.

As discussed above, reviewers must make an educated guess about the quantities given in Table 4.2. Given knowledge of the within-study information (typical sample sizes and effect size variances), the reviewer can use guesses about two of the three quantities at the level of the synthesis for a fixed effects model to solve for the third one. Before illustrating this process, I review the overall test for the mean effect size under the fixed effects model below.

4.3.1   Z-Test for the Mean Effect Size in the Fixed Effects Model

One of the first steps in a meta-analysis is to compute the mean effect size across all studies. In this section, I outline the steps of the Z-test for the mean effect size under the fixed effects model, followed by a discussion of how to conduct the power for this test. If we are interested in the effectiveness of a treatment, we may be interested in understanding if our mean effect size is different from zero, namely, if the treatment group differs significantly from the control group. In this case, our null hypothesis would be $H_0 : \theta = 0$, that our mean effect size indicates no difference between the treatment and control groups. We would use the same null hypothesis to determine if the mean correlation or log-odds ratio was different from zero. We use θ to designate the population mean, estimated by $\bar{T}_\bullet$ given in (2.1). We truly expect, however, that our treatment is effective, so that we have an alternative hypothesis that our mean effect size will be greater than 0. In this case, our alternative hypothesis is $H_a : \theta \geq 0$. This is the case of a directional alternative hypothesis which would result in a one-tailed statistical test. If we have

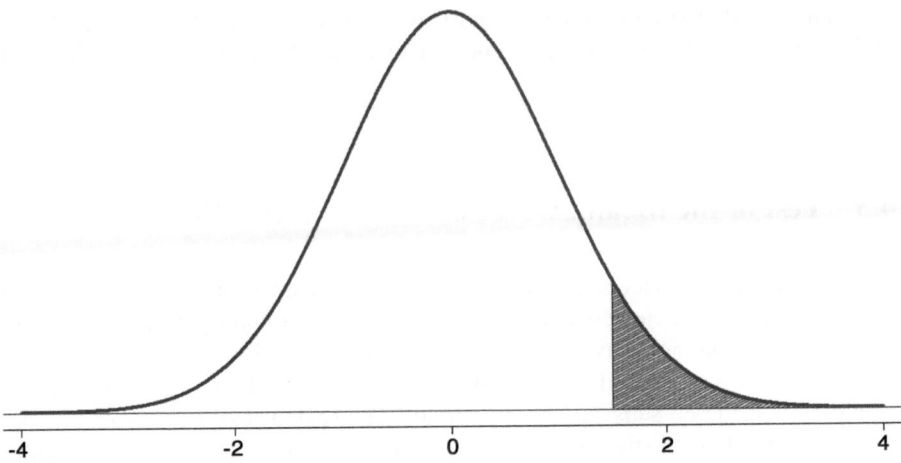

Fig. 4.1 Standard normal distribution with shaded area greater than $c_\alpha = 1.64$

little information about the intervention, we might have a less specific alternative hypothesis, allowing either the treatment or the control group to have the larger mean. This non-directional alternative hypothesis is given by $H_a : \theta \neq 0$, which results in a two-tailed test.

The test of the mean effect size compares the estimated mean effect size to 0, by comparing our obtained test statistic to a standard normal distribution. Recall from Chap. 2 that we designate the estimated effect sizes for each study as T_i, $i = 1, 2,$ ..., k where k is the total number of studies. In the fixed effects model, we have a mean effect size across all the k studies, $\bar{T}_\bullet$ given in (2.1), and this value estimates the underlying population mean effect, θ. To examine whether our obtained mean effect size, $\bar{T}_\bullet$, is statistically different from 0, we compute the following statistic

$$Z = \frac{\bar{T}_\bullet - 0}{\sqrt{v_\bullet}} \tag{4.1}$$

where $v_\bullet$ is the sampling variance of $\bar{T}_\bullet$ given in (2.2). Once we have our value of Z, we compare it to the standard normal distribution. If our alternative hypothesis is $H_a : \theta \geq \theta_a$, we reject the null hypothesis if Z is larger than the critical value, c_α, where α is our designated significance level, the $100(1 - \alpha)$ percentile of the standard normal distribution. For example, if we want to know that our mean effect size is greater than $\theta_a = 0$, our c_α for a one-tailed test with $\alpha = 0.05$ is 1.645. Figure 4.1 displays a standard normal curve. When $H_0 : \theta = 0$, the curve in Fig. 4.1 represents the distribution of our Z test statistics. The shaded part of the curve to the right is the proportion of Z statistics that exceed c_α when the null hypothesis is true.

We might also have a situation where we are not sure of the direction of the effect size, namely, that the treatment group may score higher or lower than the control group. In this situation, we have a non-directional alternative hypothesis, or,

$H_a : \theta \neq 0$. For a two-tailed test, we reject the null hypothesis $H_0 : \theta = 0$ when our obtained value of Z is either greater than or less than $c_{\alpha/2}$, or more formally when the absolute value of Z exceeds $c_{\alpha/2}$, or $Z \geq |c_{\alpha/2}|$.

4.3.2   The Power of the Test of the Mean Effect Size in Fixed Effects Models

The power of a statistical test is the probability that we will reject the null hypothesis in favor of some specific alternative hypothesis. As we see in (4.1), we need to have a value for θ, the mean effect size we are testing, and the variance of the mean effect size in a given meta-analysis in order to compute Z, and thus, in order to compute power. Below, I will discuss how we might compute these quantities prior to collecting the studies for our synthesis. For now, imagine that we do have an educated guess about these values, we have computed the statistic Z given our guesses about θ and $v_\bullet$, and we are testing that our mean effect size is greater than zero. When the null hypothesis is true, as in $H_0 : \theta = 0$, the statistic Z has the standard normal distribution with a mean of 0, and a standard deviation of 1. When the null hypothesis is false, Z has a normal distribution with a variance of 1 and a mean that is not 0 but is given by

$$\lambda = \frac{\theta - 0}{\sqrt{v_\bullet}}. \tag{4.2}$$

where the value θ is our mean effect size, and $v_\bullet$ is the sampling variance of this mean effect size. Thus, we have two different normal distributions we are going to compare: the standard normal distribution that corresponds to the null hypothesis, and the normal distribution that represents our alternative hypothesis. To compute power, we need to know the proportion of test statistics that exceed c_α in the normal distribution for the alternative hypothesis with mean λ from (4.2), and variance 1. We get this proportion, the power of our test, by computing p

$$p = 1 - \Phi(c_\alpha - \lambda) \tag{4.3}$$

where $\Phi(x)$ is the cumulative distribution function of the standard normal distribution, i.e., the area under the standard normal curve from $-\infty$ to x. For example, Fig. 4.2 shows both the standard normal distribution and the normal distribution of Z test statistics when $\theta \neq 0$, in this illustration when $\theta = 2.5$. The proportion of tests in the cross-hatched area is equal to the area under the standard normal curve that exceeds the value of $c_\alpha - \lambda$ when $C_\alpha = 1.64$ and $\lambda = 2.50$. The exact value of this area is $1 - \Phi(C_\alpha - \lambda) = 1 - \Phi(-0.86) = 1 - 0.19 = 0.81$. Thus, the power of the test under this alternative hypothesis, $H_a : \theta \geq 0$, is 0.81.

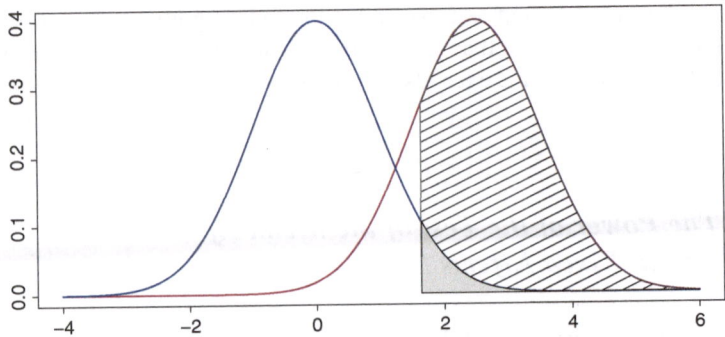

Fig. 4.2 Power when normal distribution has a mean of 2.5 and $c_\alpha = 1.64$

A two-tailed test follows similar logic, except that we are interested in the proportion of test statistics that are either greater than or less than $c_{\alpha/2}$, or $Z \geq |c_{\alpha/2}|$. The test statistics are in the two tails of the distribution so that the power for our two-tailed test is given by

$$p = 1 - \left[\Phi(c_{\alpha/2} - \lambda) - \Phi(-c_{\alpha/2} - \lambda) \right]$$
$$= 1 - \Phi(c_{\alpha/2} - \lambda) + \Phi(-c_{\alpha/2} - \lambda) \tag{4.4}$$

4.3.3   Deciding on Values for Parameters to Compute Power

In order to compute power of the mean effect size, we need to designate a number of quantities. Our first task is to arrive at a value for θ, the mean effect size of interest, or, in other words, the value of a substantively important effect. This value should be evident from the context of the review and the typical measures used. For example, we may know that increasing a child's reading scores from the 25th to the 50th percentile is associated with higher achievement in later elementary grades. We would then be interested in the power of a test to detect an effect size corresponding to an increase of 25 percentile points. As seen in (4.2), we also need an estimate of the variance for the mean effect size of interest. This quantity is more difficult to determine. Looking at (2.2), we see that the variance for the mean effect size, $v_\bullet$, depends on the variances for the effect sizes within each study, which in turn depend on the individual study effect sizes and the within-study sample sizes. One way to think about an estimate for $v_\bullet$ is to assume that all studies have the same within-study sample size, and the same effect size equal to θ, our effect size of interest. With the same within-study sample size and effect size, our studies would then all have the same within-study effect size variance, v. Note that we compute $v_\bullet$ in this case as

$$v_\bullet = \frac{1}{\sum_{i=1}^{k} w} = \frac{1}{\sum_{i=1}^{k} 1/v} = \frac{1}{k/v} = \frac{v}{k}. \qquad (4.5)$$

Finally, we need a value of k, the number of studies we are likely to find as relevant in the review. In sum, to compute the power for the mean effect size in a fixed effects model, we need a value for the effect size of interest, θ, the within-study sample size (to compute v), and the total number of studies, k. Below I provide a number of examples.

4.3.4   Example: Computing the Power of the Test of the Mean

One of the examples we will be examining throughout the text is a meta-analysis of studies on the effectiveness of coaching programs to increase the SAT-Math and SAT-Verbal scores of high school students. (The SAT is a common standardized test used by US colleges and universities in their enrollment decisions). If I am a parent of a high school student, I would be interested in investing in a coaching program if my child had the potential to increase their score by 50 points. Since the SAT test has standard deviation of 100 points, I am interested in an effect size of $\theta = 50/100 = 0.5$. To compute power for this effect size, I also need to have a guess about the number of studies that I will include in the synthesis and the typical sample size within those studies in order to estimate the variance of the mean effect size. Let us say that the typical study has a total of 40 participants, 20 in the coaching group and 20 in the control. Let us also guess that we will find 10 studies that are relevant to our review.

Given these numbers, we can compute the common within-study variance, v, using (2.13), as

$$v = \frac{20 + 20}{20 * 20} + \frac{0.5^2}{2(20 + 20)} = 0.10.$$

With $k = 10$ studies, we have a value of $v_\bullet = (0.10/10) = 0.01$ using (4.5). We can now compute the power for our test given an effect size of interest equal to 0.5, by first computing λ, the mean for our alternative hypothesis,

$$\lambda = \frac{0.5 - 0}{\sqrt{0.01}} = 5.$$

In this case, we are interested in whether the coaching group outperforms the control group as represented by a mean effect size of 5.0, so we will compute the power for a one-tailed test as given in (4.3). For $\alpha = 0.05$, we have power of $p = 1 - \Phi(1.645 - 5.0) = 1 - \Phi(-3.36) = 1$. Thus, we have sufficient power to find an effect size of 0.5 with only 10 studies that have a sample size of 40.

Let us keep the same number of studies, but decrease both the within-study sample sizes and the effect size of interest to examine how the power changes. Let us say that we expect to find 10 studies, but the within-study sample size is now $N = 20$, with 10 participants in each group. In addition, we want to know the power for an effect size of 0.20, which corresponds to a 20 point difference between the coaching and control groups since the standard deviation of the SAT is 100 points. In this case, let us also use a non-directional test – either the control group or the treatment group may obtain the higher SAT score. With a total $N = 20$, with 10 in each group, we would compute a value of $v = 0.201$. If we have $k = 10$ studies, then

$$v_\bullet = 1/(\sum 1/v_i) = 1/(10/0.201) = 0.02$$

The value of λ in the SAT coaching study is given by

$$\lambda = \frac{0.20 - 0.00}{\sqrt{0.02}} = \frac{0.20}{0.14} = 1.41$$

The next step is to determine the proportion of Z test statistics that are either greater or less than a given critical value in a normal distribution with mean 1.41 and variance 1. If we are interested in a two-tailed test with $\alpha = 0.05$, then the critical value $c_{\alpha/2}$ is equal to the 97.5 percentile of the standard normal distribution, or 1.96. In the other tail, we are interested in the critical value equal to the 2.25 percentile or -1.96. We obtain the proportion of Z-test statistics above 1.96 and below -1.96 by using the cumulative normal distribution function. To compute this value, we use (4.4) and obtain test statistics that are greater than a value of $1.96 - (1.41) = 0.55$ and that are less than $-1.96 - (1.41) = -3.37$ in the standard normal distribution.

The power of the two-sided test for the standardized mean difference is then computed as

$$\begin{aligned} p &= 1 - [\Phi(1.96 - 1.41) - \Phi(-1.96 - 1.41)] \\ &= 1 - \Phi(0.55) + \Phi(-3.37) \\ &= 1 - 0.71 + 0 \\ &= 0.29 \end{aligned}$$

We have low power to find that an effect size of 0.20 is significantly different from zero, or correspondingly, to detect a test-score difference of 20 points when we assume that our studies have total samples sizes of $N = 20$. Figure 4.3 shows the power for the two-sided test in this example; note that only the value to the right of the standard normal curve is nonzero. The Appendix to this chapter provides

Fig. 4.3 Power when normal
distribution has a mean of
1.41 and $c_\alpha = 1.96$

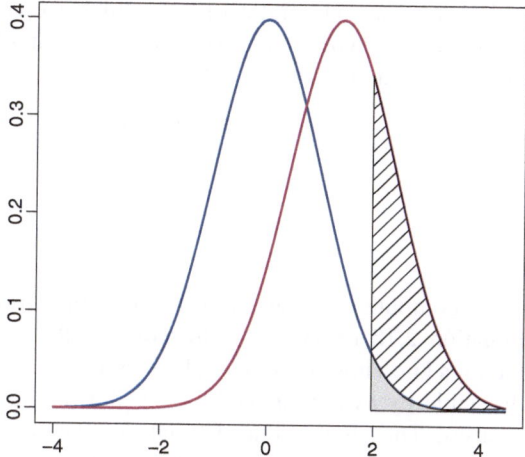

information about how to compute the values of the cumulative normal distribution
in Excel, SPSS, and R.

4.3.5   Example: Computing the Number of Studies Needed to Detect an Important Fixed Effects Mean

The National Research Council (1986) examined a number of studies of the
effects of exposure to secondhand smoke, and found that non-smoking spouses
exposed to tobacco smoke had a higher risk of developing lung cancer than
non-exposed spouses. This report estimated an odds ratio of 1.30 for developing
lung cancer in exposed spouses versus non-exposed spouses from a set of US
studies completed at the time of the report. If we were planning to update this
review, we might want to know how many studies we need to find an odds ratio
of 1.30 as statistically significant, with power of .80. Estimating the number of
studies needed before collecting the data will help both to guide and to interpret
our findings.

The number of studies needed, k, relates to power through the estimate of our the
common variance of our effect size, $v_\bullet$, as given in (4.5). We will work backwards,
given our value for power of 0.8 and a significance level, α, of 0.05, first obtaining a
value for λ, and then solving for $v_\bullet$ given an odds ratio of 1.30. To arrive at power
0.8 in a one-tailed test with $\alpha = 0.05$, we solve the following

$$1 - \Phi(1.645 - \lambda) = 0.8$$
$$\Phi(1.645 - \lambda) = 0.2$$
$$1.645 - \lambda = -0.842$$
$$\lambda = 2.487$$

We have an odds ratio of 1.3, which we convert to a log-odds ratio for our computations, $LOR = \ln(1.3) = 0.252$. Solving for $v_\bullet$ given our value of λ and our log-odds ratio gives us

$$\frac{0.252 - 0}{\sqrt{v_\bullet}} = 2.487$$

$$\sqrt{v_\bullet} = 0.101$$

$$v_\bullet = 0.010$$

To compute k, we need to get an estimate of the common within-study variance. Equation 2.18 provides the formula for the variance of a log-odds ratio, which requires knowing the actual values in the 2×2 table of results. We can reproduce that table by assuming sample sizes for the number of exposed and non-exposed spouses. The studies that estimate of the risk of lung cancer with secondhand smoking exposure are usually large epidemiological studies. Let us assume that there are a total of 2,000 cases, with 1,000 cases within each group. In the general population, the current probability of contracting lung cancer between the ages of 50 and 70 is 2.89% for men and 2.27% for women (National Cancer Institute 2011). If we take a conservative control base rate of 2%, with 1,000 cases for both exposed and non-exposed spouses, we could create a 2×2 table as seen in Table 4.3.

Table 4.3 Hypothetical lung cancer incidence rate for exposure to secondhand smoke

Group	Lung cancer	No lung cancer
Exposed spouses	26	974
Not exposed spouses	20	980

The table above results in an odds ratio of 1.31, and a variance of the log-odds ration of

$$v = \frac{1}{26} + \frac{1}{974} + \frac{1}{20} + \frac{1}{980}$$

$$v = 0.09$$

Given this value of v, we can now solve for k using (4.5), where $v_\bullet = v/k$, or in our case, $0.01 = 0.09/k$ so that $k = 9$. Thus, if our search finds at least 9 studies that have 2,000 participants within each study, we will have sufficient power to detect an odds ratio of 1.3.

4.3.6   Example: Computing the Detectable Fixed Effects Mean in a Meta-analysis

We might also be interested in knowing the size of the effect that would be considered statistically significant (as opposed to substantively significant) given

k, the number of studies relevant to the review, a particular value of power, and typical within-study sample sizes. Note that this computation will be possible with the odds-ratio and correlation as effect sizes since the within-study variance of the effect size for these two indices depends only on the sample size. For the standardized mean difference, it will be more difficult to solve for the detectable effect size given that the within-study variance depends on the effect size itself. For this example, suppose we have a set of studies that examines the relationship between student evaluations of instructors and student course grades. Let us assume that we will find $k = 15$ studies, each with a within-study sample size of 20. For power of 0.80, we can solve for the effect size that we will be able to find statistically different from zero. Using (4.3), we could solve for the value of λ that will produce a power of 0.8 as we did in Example 4.3.4, obtaining a value of $\lambda = 2.487$. We need the variance of the effect size given in (2.15), noting that we typically use Fisher's z-transformation instead of the correlation. For Fisher's z-transformation, we would compute the within-study variance is given as $v = 1/(n - 3) = 1/(20 - 3) = 1/17 = 0.059$ when $N = 20$. Our estimated value for the variance of the mean effect is found in (4.5), resulting in $v_\bullet = 0.059/15 = 0.004$. We can now solve for θ, as

$$\frac{\theta - 0}{\sqrt{0.004}} = 2.487$$

$$\frac{\theta}{0.063} = 2.487$$

$$\theta = 0.16$$

Our estimate of the effect size is given as Fisher's z-transformation. Transforming back to the correlation metric using (2.16) gives us an estimated $r = 0.16$. Thus, with $k = 15$ studies, each with a sample size of 20, we can find a correlation equal to 0.16 as significantly different from zero with power 0.80.

4.4 Test of the Mean Effect Size in the Random Effects Model

Recall that in the random effects model, we are assuming that the variation among studies' effect sizes has two components, one due to sampling variance designated by v_i, and one due to the variance in the underlying distribution of effect sizes, or τ^2. As we see in Table 4.2, the power of the test of the mean effect size in a random effects model depends on one additional parameter from the fixed effects model, the random effects variance, τ^2. In order to compute power for the random effects mean, we will need to have an educated guess about the value of τ^2. Below I provide the details for the power of the random effects mean effect size, and then discuss how we might arrive at a value for τ^2.

4.4.1 The Power of the Test of the Mean Effect Size in Random Effects Models

The test of the mean effect size in the random effects model takes a form similar to the test for the fixed effects mean. We will again compare our estimated mean effect size to 0, by comparing our obtained test statistic to a standard normal distribution. To examine whether our obtained mean effect size, $\bar{T}_{\bullet}^*$ (as given in (2.7)) is statistically different from 0, we compute the following statistic

$$Z^* = \frac{\bar{T}_{\bullet}^* - 0}{\sqrt{v_{\bullet}^*}} \tag{4.6}$$

where $v_{\bullet}^*$ is the variance of $\bar{T}_{\bullet}^*$ given in (2.8). Once we have our value of Z^*, we compare it to the standard normal distribution.

As I discuss above, the power of a statistical test is the probability that we will reject the null hypothesis in favor of some specific alternative hypothesis. To compute power, we need to have a value for θ^*, the random mean effect size we are testing, and the random effects variance of the mean effect size in a given meta-analysis in order to compute Z^*. Below, I will discuss how we might compute a value for the random effects variance of the mean effect size prior to collecting the studies for our synthesis. For now, imagine that we have computed the statistic Z^* given our guesses about θ^* and $v_{\bullet}^*$, and we are testing that our mean effect size is greater than zero. When the null hypothesis is true, as in $H_0 : \theta^* = 0$, the statistic Z^* has the standard normal distribution with a mean of 0, and a standard deviation of 1. When the null hypothesis is false, Z^* has a normal distribution with a variance of 1 and a mean that is not 0 but is given by

$$\lambda^* = \frac{\theta^* - 0}{\sqrt{v_{\bullet}^*}}. \tag{4.7}$$

where θ^* is our random effects mean effect size, and $v_{\bullet}^*$ is the variance of our random effects mean. To compute power, we need to know the proportion of test statistics that exceed c_α in the normal distribution for the alternative hypothesis with mean λ^* from (4.7), and variance 1. We get this proportion, the power of our test, by computing p

$$p = 1 - \Phi(c_\alpha - \lambda^*) \tag{4.8}$$

where $\Phi(x)$ is the cumulative distribution function of the standard normal distribution, i.e., the area under the standard normal curve from $-\infty$ to x.

A two-tailed test follows similar logic, except that we are interested in the proportion of test statistics that are either greater than or less than $c_{\alpha/2}$, or $Z^* \geq |c_{\alpha/2}|$.

The test statistics are in the two tails of the distribution so that the power for our two-tailed test is given by

$$p = 1 - \left[\Phi(c_{\alpha/2} - \lambda^*) - \Phi(-c_{\alpha/2} - \lambda^*)\right]$$
$$= 1 - \Phi(c_{\alpha/2} - \lambda^*) + \Phi(-c_{\alpha/2} - \lambda^*) \tag{4.9}$$

4.4.2 Positing a Value for τ^2 for Power Computations in the Random Effects Model

The power of the random effects mean requires us to posit a value of the mean effect size, θ^*, the number of studies, k, and a value for λ^*, which depends on the variance of the random effects mean, $v_\bullet^*$. In the section on the power of test of the fixed effects mean, I illustrate ways we could think about important values for θ, k, and v, the common effect size variance. An added complication in the test of the random effects mean is the need to have a guess about the value of the variance component τ^2 in order to compute the variance of the random effects mean.

In Sect. 4.3.3, I suggest that we assume that all effect sizes have the same within-study variance, or v, in the fixed effects model. This simplification allows us to estimate the variance of the fixed effects mean equal to v/k. Since the random effects variance of an individual effect size is equal to $v_i^* = v + \tau^2$, we can use our convention for assuming a common effect size variance v for part of our estimate of the random effects variance.

For estimating a value of τ^2, we can adopt conventions suggested by Higgins and Thompson (2002). Higgins and Thompson's paper introduces another index of heterogeneity, I^2, that is interpreted as the percent of variation that is due to heterogeneity among effect sizes rather than sampling variance. Higgins and Thompson consider values of I^2 equal to 25%, 50%, and 75% as small, moderate and large degrees of heterogeneity. We can write I^2 in terms of a target value of the variance component, τ^2, and a "typical" value for the within-study effect size variance that we have been denoting as v, or,

$$I^2 = \frac{\tau^2}{\tau^2 + v} \tag{4.10}$$

We can then posit values of I^2 that could be plausible for a given set of studies. If I^2 is equal to .75, a large degree of heterogeneity, then we could solve (4.10) as

$$\frac{\tau^2}{\tau^2 + v} = .75$$
$$\tau^2 = .75(\tau^2 + v)$$
$$.25\tau^2 = .75v$$
$$\tau^2 = 3v$$

For a moderate degree of heterogeneity, $\tau^2 = v$. A low degree of heterogeneity would correspond to $\tau^2 = (1/3)v$. Given a value of v, we can obtain an estimate for τ^2, and thus can compute our value for $v_\bullet^*$ as

$$v_\bullet^* = \frac{1}{\displaystyle\sum_{i=1}^{k} w_\bullet^*} = \frac{1}{\displaystyle\sum_{i=1}^{k} 1/(v + \tau^2)} = \frac{1}{k/(v + \tau^2)} = \frac{v + \tau^2}{k} \qquad (4.11)$$

4.4.3   Example: Estimating the Power of the Random Effects Mean

In Example 4.3.4, we estimated the power of the fixed effects mean for a hypothetical set of studies on the effects of coaching on SAT scores. Recall that we wanted to be able to detect an effect size of $\theta = 50/100 = 0.5$. We also assumed that the typical study has a total of 40 participants, 20 in the coaching group and 20 in the control, and that we will find 10 studies that are relevant to our review. We also assumed that the coaching group would score higher than the control group, leading us to use a one-tailed test with $\alpha = 0.5$. Given these assumptions, we find a common within-study effect size variance of $v = 0.10$. For the random effects mean, we also need an estimate of the random effects variance, or, τ^2. Let us compute power for two different values of τ^2, one assuming a large degree of heterogeneity, and one assuming a small degree of heterogeneity. With our value of v, we can compute for a large degree of heterogeneity, $\tau^2 = 3v = 0.30$, and for a small degree of heterogeneity, $\tau^2 = (1/3)v = (1/3)0.10 = 0.033$. With $k = 10$ studies and a large degree of heterogeneity, we have $v_\bullet^* = (0.10 + 0.30)/10 = 0.04$ as given by (4.11). For a small degree of heterogeneity, we have $v_\bullet^* = (0.10 + 0.033)/10 = 0.133/10 = 0.0133$. A large degree of heterogeneity gives us $\lambda = (0.5 - 0)/\sqrt{0.04} = 2.5$, and power of $1 - \Phi(1.645 - 2.5) = 1 - \Phi(-0.855) = 1 - 0.20 = 0.80$. A small degree of heterogeneity gives us $\lambda = (0.5 - 0)/\sqrt{0.0133} = 4.34$ and power of $1 - \Phi(1.645 - 4.34) = 1 - \Phi(-2.69) = 1 - 0.003 = 0.997$. When we have a large degree of heterogeneity, we have less power to detect an effect size of 0.5 since we have more variability among our effect size estimates than when we have a small degree of heterogeneity.

Now let us compute power for two values of τ^2 when we assume a smaller effect size, $\theta = 0.2$, within-study sample size $N = 20$, or 10 for each group, and a two-tailed test with $\alpha = 0.05$. As in the second half of Example 4.3.4, with the values above, we obtain an estimate of the common within-study variance, v, equal to 0.20. For a large degree of heterogeneity, our $\tau^2 = 3v = 3 * 0.20 = 0.6$. A small degree of heterogeneity results in $\tau^2 = v/3 = 0.20/3 = 0.07$. With $k = 10$ studies and a large degree of heterogeneity, we have $v_{\bullet}^* = (0.20 + 0.6)/10 = 0.08$ as given by (4.11). For a small degree of heterogeneity, we have $v_{\bullet}^* = (0.20 + 0.07)/10 = 0.27/10 = 0.027$. A large degree of heterogeneity gives us $\lambda = (0.2 - 0)/\sqrt{0.08} = 0.71$. If we are interested in a non-directional test, the power for a large degree of heterogeneity is $1 - \Phi(1.96 - 0.71) + \Phi(-1.96 - 0.71) = 1 - \Phi(1.25) + \Phi(-2.67) = 1 - 0.89 + 0 = 0.11$. A small degree of heterogeneity gives us $\lambda = (0.2 - 0)/\sqrt{0.027} = 1.22$ and power of $1 - \Phi(1.96 - 1.22) + \Phi(-1.96 - 1.22) = 1 - \Phi(0.74) + \Phi(-3.18) = 1 - 0.77 + 0 = 0.23$. In general, the power is lower for the test of the random effects mean as we see in the example above compared to Example 4.3.4. In addition, larger degrees of heterogeneity result in less powerful tests of the mean effect size.

4.4.4   Example: Computing the Number of Studies Needed to Detect an Important Random Effect Mean

In Example 4.3.5, we computed the number of studies needed to detect an odds ratio of 1.3 for the non-smoking spouse of a smoker compared to the non-smoking spouse of a non-smoker of contracting lung cancer. We can use the results we obtained in this example to find the number of studies needed to detect this odds ratio given a large and a small degree of heterogeneity. In Example 4.3.5, we found a value of $\lambda = 2.487$. Using the same reasoning, we then would obtain a value for our $v_i^* = 0.010$. If we use the hypothetical data given in Table 4.3 for an odds ratio of 1.31, we obtain a common value of $v = 0.09$. Given v_i^*, v and a value for τ^2, we can solve for k. A large degree of heterogeneity would give us $\tau^2 = 3(0.09) = 0.27$. We then must solve

$$0.01 = \frac{(0.09 + 0.27)}{k}$$
$$k = \frac{0.36}{0.01}$$
$$k = 36$$

For a small degree of heterogeneity, we obtain $\tau^2 = (1/3)0.09 = 0.03$, and solving for k given the values above yields $k = 12$. Note that we need more studies to detect an odds ratio of 1.3 when we assume a random effects model, and more studies still when we have more heterogeneity among effect sizes.

4.4.5   Example: Computing the Detectable Random Effects Mean in a Meta-analysis

Extending example 4.3.6, we can compute the size of the random effects mean correlation that we would find statistically significant given k, a value of power, the typical within-study variance and a value for τ^2. In example 4.3.6, we assumed $k = 15$, and the common within-study variance, v, for Fisher's z-transformation as 0.059 (for $N = 20$) for the association between instructor ratings and student grades. For a large degree of heterogeneity, we have $\tau^2 = 3(0.059) = 0.18$, and for a small degree of heterogeneity, we have $\tau^2 = (1/3)0.059 = 0.02$. For a one-tailed test with power of 0.8, we have $\lambda = 2.487$. Our corresponding values for $v_i^* = (0.059 + 0.18)/15 = 0.016$ for a large degree of heterogeneity, and $v_i^* = (0.059 + 0.02)/15 = 0.005$ for a small degree. We solve for θ^* with a large degree of heterogeneity as

$$\frac{\theta^* - 0}{\sqrt{0.016}} = 2.487$$

$$\frac{\theta^*}{0.13} = 2.487$$

$$\theta^* = 0.32$$

With a small degree of heterogeneity, we obtain $\theta^* = 0.19$. Both values are larger than the effect we can detect as statistically significant when we assume a fixed effects model.

Appendix

Computing Power for Examples in Section 4.3

The examples require the computation of the area under the standard normal distribution that is either less than or greater than a given critical value, c_α. Below I give the functions and commands necessary to obtain the power for the example in Sect. 4.3.4.

Excel

In Excel, the function NORMSDIST(x) is the cumulative normal distribution. For Example 4.3.4,

$$\text{NORMSDIST}(0.55) = 0.709$$

The value given in Excel is the cumulative area less than or equal to x.

SPSS

Using the Compute menu in SPSS, the function CDF.NORMAL(quant, mean, sd) provides the values for the cumulative normal distribution. To compute the power for Example 4.3.4,

$$\text{value} = \text{CDF.NORMAL}(0.55, 0, 1)$$

returns the value 0.71, or in other words, the cumulative area that is less than or equal to 0.71.

SAS

In SAS, the function CDF('NORMAL', x, mean, sd) provides the value of the cumulative normal distribution function that is less than x for a normal distribution with the specified mean and standard deviation. To compute the power for Example 4.3.1,

$$\text{value} = \text{CDF('NORMAL'}, 0.55, 0, 1).$$

The above function results in a value of 0.709.

R

In R, the function PNORM(x) gives the area for the cumulative standard normal distribution to the right of a positive value of x and to the left of a negative value of x. For Example 4.3.4, the following command produces the area that is greater than x (since x is positive), or, $P(X > x)$. The command pnorm(0.55) in R would produce the same result as we see below.

$$> \text{pnorm}(0.55)$$
$$> 0.709$$

References

American Psychological Association. 2009. *Publication manual of the American Psychological Association*, 6th ed. Washington, DC: American Psychological Association.

Cohen, J. 1988. *Statistical power analysis for the behavioral sciences*, 2nd ed. Mahwah: Lawrence Erlbaum.

Freiman, J.A., T.C. Chalmers, H. Smith, and R.R. Kuebler. 1978. The importance of beta, the type II error and sample size in the design and interpretations of the randomized control trial. *The New England Journal of Medicine* 229: 690–694.

Goodman, S.N., and J. Berlin. 1994. The use of predicted confidence intervals when planning experiments and the misuse of power when interpreting results. *Annals of Internal Medicine* 121: 201–206.

Hayes, J.P., and R.J. Steidl. 1997. Statistical power analysis and amphibian population trends. *Conservation Biology* 11: 273–275.

Higgins, J.P.T., and S.G. Thompson. 2002. Quantifying heterogeneity in a meta-analysis. *Statistics in Medicine* 21: 1539–1558.

Murphy, K.R., and B. Myors. 2004. *Statistical power analysis*, 2nd ed. Mahwah: Lawrence Erlbaum.

National Cancer Institute. 2011. Cancer fact sheet: Cancer of the lung and bronchus. National Institutes of Health. http://seer.cancer.gov/statfacts/html/lungb.html. Accessed 25 July 2011.

National Research Council. 1986. *Environmental tobacco smoke: Measuring exposures and assessing health effects*. Washington, DC: National Academy Press.

Reed, J.M., and A.R. Blaustein. 1995. Assessment of "nondeclining" amphibian populations using power analysis. *Conservation Biology* 9: 1299–1300.

Thomas, L. 1997. Retrospective power analysis. *Conservation Biology* 11: 276–280.

Zumbo, B.D., and A.M. Hubley. 1998. A note on misconceptions concerning prospective and retrospective power. *The Statistician* 47: 385–388.

Chapter 5
Power for the Test of Homogeneity in Fixed and Random Effects Models

Abstract This chapter will illustrate methods for the power of the test of homogeneity in fixed and random effects models. In fixed effects models, the test of homogeneity provides evidence about whether the effect sizes in a meta-analysis are measuring a common effect size. The test of homogeneity in random effects models is a test of the statistical significance of the variance component, the between-studies variance. The chapter gives examples of how to compute the power for the test of homogeneity in both fixed and random effects models.

5.1 Background

The prior chapter provided background on power analysis in general, and outlined the specific tests for the power of the mean effect size in both fixed and random effects models. This chapter will discuss the power for tests of homogeneity of effect sizes in both fixed and random effects models. For fixed effects, the test of homogeneity examines whether the amount of variation among effect sizes is greater than we would expect due to sampling error alone. For random effects models, the test of homogeneity is the test of whether the variance component, τ^2, is different from zero. Though each of these tests uses the same computational form, their underlying distribution is different for the computation of power.

As seen in the prior chapter, one challenge in examining power is deciding on the values to use for computing power. The test of homogeneity requires an a priori guess about the amount of variation that we expect among effect sizes in a meta-analysis. While we often have a priori ideas about values for the mean effect size, it is more difficult to quantify an amount of heterogeneity. I present two options in this chapter, one by thinking about the average amount of difference between the effect sizes and the mean effect in standard deviation units, and one by using percentages of I^2. Below I provide the power for the tests of homogeneity in fixed and random effects models, and suggest ways to choose values for the parameters needed to compute power.

T.D. Pigott, *Advances in Meta-Analysis*, Statistics for Social and Behavioral Sciences, DOI 10.1007/978-1-4614-2278-5_5, © Springer Science+Business Media, LLC 2012

5.2   The Test of Homogeneity of Effect Sizes in a Fixed Effects Model

In Chap. 3, I discussed the assumptions we make in fixed and random effects models. The major distinction between these models lies in the source of the variation among effect size estimates. For a fixed effects model, we assume that the effect size estimates from studies differ due only to sampling error. Each study is estimating a common mean effect size, and the estimates differ from each other since each study uses a different sample of participants. We test this assumption using the homogeneity statistic, Q. In a meta-analysis, after we estimate our overall fixed effects mean effect size, we are interested in testing the null hypothesis that all studies estimate a common effect size, or $H_0 : \theta_1 = \theta_2 = \ldots = \theta_k = \theta$. This test was given in (2.3), and can be written as

$$Q = \sum_{i=1}^{k} \frac{(T_i - \bar{T}_{\bullet})^2}{v_i} = \sum_{i=1}^{k} w_i (T_i - \bar{T}_{\bullet})^2$$

where $w_i = 1/v_i$, the inverse of the sampling variance of the effect size estimates, T_i, and the mean of the effect size estimates, $\bar{T}_{\bullet}$, is given by (2.1). We use θ to designate the common population mean estimated by $\bar{T}_{\bullet}$.

We compare the obtained value of Q to a chi-square distribution with $k - 1$ degrees of freedom. We reject the null hypothesis if the value of Q is larger than the critical value, c_α, where α is our designated significance level, the $100(1 - \alpha)$ percentile of the chi-square distribution with $k - 1$ degrees of freedom. The test of homogeneity is a one-tailed test where we reject H_0 when $Q \geq c_\alpha$. Note that a nonsignificant value of the chi-square distribution indicates that our effect sizes are homogeneous, that they are estimating a single common mean effect size.

5.2.1   The Power of the Test of Homogeneity in a Fixed Effects Model

In order to compute the power of the test of homogeneity in fixed effects models, we will need to have an estimate of Q. Below I discuss how we might posit values for the homogeneity statistic. For now, imagine that we are able to arrive at a value for Q. As indicated above, when the null hypothesis is true, i.e., when all effect sizes estimate a common mean, Q has the chi-square distribution with $k - 1$ degrees of freedom. When at least one of the effect sizes differs, i.e., when the alternative hypothesis, H_a: $\theta_i \neq \theta$ for some value of i, then Q has a non-central chi-square distribution with $k - 1$ degrees of freedom and non-centrality parameter equal to

$$\lambda = \sum_{i=1}^{k} w_i (\theta_i - \bar{\theta}_\bullet)^2. \tag{5.1}$$

where $\bar{\theta}_\bullet$ is the weighted mean of $\theta_1, \theta_2, \ldots, \theta_k$, and can be written as

$$\bar{\theta}_\bullet = \frac{\sum_{i=1}^{k} w_i \theta_i}{\sum_{i=1}^{k} w_i}. \tag{5.2}$$

As in the power of the test of the mean effect size, we want to know the proportion of Q statistics that are greater than c_α when we have a non-central chi-square distribution with a non-centrality parameter λ and $k - 1$ degrees of freedom. Note that the value of c_α is the critical value from a central chi-square distribution with $k - 1$ degrees of freedom. Note that (5.1) has the same form as the homogeneity test, Q. We will propose a value of Q to estimate the non-centrality parameter, λ, when we compute power. The power of the test of homogeneity in the fixed effects case is given by

$$p = 1 - F(c_\alpha \mid k - 1; \lambda) \tag{5.3}$$

where $F(c_\alpha \mid k - 1; \lambda)$ is the area that is larger than c_α of the non-central chi-square with $k - 1$ degrees of freedom and non-centrality parameter λ.

5.2.2   Choosing Values for the Parameters Needed to Compute Power of the Homogeneity Test in Fixed Effects Models

The non-centrality parameter is key in computing power for the homogeneity of effect sizes in a fixed effects context. Looking at (5.1), we see that one component of the non-centrality parameter is the difference between each effect size, θ_i, and the mean effect size θ. We could think about this difference in terms of the units of the standard deviation of the mean effect size, $\sqrt{v_\bullet}$. We might want to see if we could detect heterogeneity if the average difference between the effect sizes and the mean effect was $.5\sqrt{v_\bullet}$, one-half of a standard deviation of the mean effect size. If we also have assumptions about the typical sample size within studies, N, and the number of studies we expect to find, k, we can compute a value for the common variance for each effect size, and the estimated variance for the mean of those effect sizes as we did in Chap. 4. We can then compute a non-centrality parameter λ, in this example, as

$$
\begin{aligned}
\lambda &= \sum_{i=1}^{k} w \left(0.5 \sqrt{v_\bullet}\right)^2 \\
&= k \, w \, v_\bullet (0.5)^2 \\
&= \frac{k \, v_\bullet (0.5)^2}{v} \\
&= \frac{k}{v} \frac{v}{k} (0.5)^2 \\
&= (0.5)^2
\end{aligned}
\tag{5.4}
$$

In (5.4), the estimate of $v_\bullet$ is given in (2.2), and $w = 1/v$ is the inverse of the common variance for each effect size. Given that we estimate $v_\bullet = v/k$, one estimate of λ is equal to the square of the number of standard deviations we propose as the average difference between effect sizes and their mean. We could consider other values for the average difference between each study's effect size and the mean effect size such as $1v_\bullet$ or $2v_\bullet$, resulting in a value of λ as 1, or 4, respectively. Below I provide an example of these computations.

5.2.3   Example: Estimating the Power of the Test of Homogeneity in Fixed Effects Models

In Chap. 4, we used an example of studies on the effectiveness of coaching programs to increase the SAT-Math and SAT-Verbal scores of high school students. In that example, we were interested in a standardized mean difference of 0.5, and assumed that a typical study had a total sample size of 40, with 20 participants each in the experimental and control groups. In this example, we obtained a value of the common effect size of $v = 0.10$. With $k = 10$ studies, we have $v_\bullet = (0.10/10) = 0.01$. If we assume that the average difference among the effect sizes and the mean effect size is $0.5 \sqrt{v_\bullet}$, then we would compute a value of the non-centrality parameter equal to

$$
\lambda = (0.5)^2 = 0.25
$$

The critical value of the central chi-square is 16.92 for a central chi-square with $k - 1 = 9$ degrees of freedom and $\alpha = 0.05$ meaning that 95% of the central chi-square distribution with 9 degrees of freedom lies below the value of 16.92. The power of the homogeneity test in this case is $p = 1 - F(c_\alpha | k - 1; \lambda) = 1 - F \times (16.92 | 9; 0.25) = 0.058$. We do not have much power to detect heterogeneity when the average difference between effect sizes and the mean is only $0.5 \sqrt{v_\bullet}$. If we

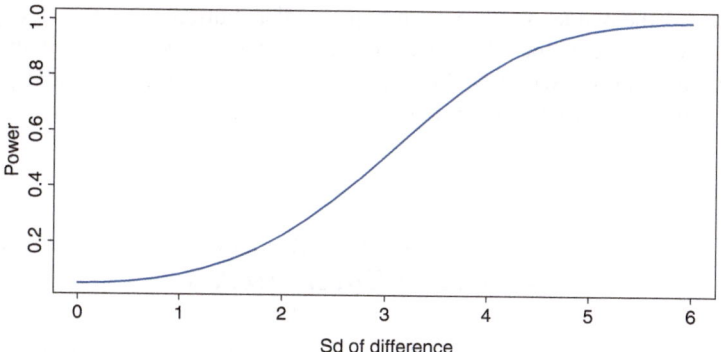

Fig. 5.1 Power for differences between the effect size estimates and the mean in SD units

assume that the average difference between effect sizes and the mean is larger, say $2v_{\bullet}$, we would obtain a value of the non-centrality parameter equal to $\lambda = 4$. The power of the homogeneity test is $p = 1 - F(c_\alpha \,|\, k - 1;\, \lambda) = 1 - F(16.92 | 9; 4) = 0.22$. When we have more heterogeneity, say when the average difference between the effect sizes and the mean is $3\sqrt{v_{\bullet}}$, the power is $p = 1 - F(c_\alpha \,|\, k - 1;\, \lambda) = 1 - F(16.92 | 9; 9) = 0.51$. Figure 5.1 provides a graph of the power for this example using different values for the average difference between the effect sizes and the mean in standard deviations from 0 to 6. Note that our power to detect heterogeneity reaches 0.8 only when our effect sizes are on average 4 standard deviations away from the mean effect size.

5.3   The Test of the Significance of the Variance Component in Random Effects Models

As discussed in Chap. 3, we assume that the variation between effect sizes in a random effects model consists of two components: one due to sampling error, and one to the variation in the underlying distribution of effect sizes. When we are testing homogeneity in the random effects model, we are interested in whether the underlying variation in the distribution of effect sizes is zero, i.e., whether the only source of variation is credibly sampling error. Thus, the test of homogeneity in the random effects model is a test of whether the variance component, τ^2, is equal to 0. Thus, the null hypothesis for the test of the variance component in random effects models is $H_0\colon \tau^2 = 0$, with the alternative hypothesis of $H_a\colon \tau^2 \neq 0$.

The test that the variance component is equal to zero, $\tau^2 = 0$, in the random-effects model has a form that is similar to the homogeneity test, Q, in the fixed effects case. As in (2.9), the test is given as,

$$Q = \sum_{i=1}^{k} w_i (T_i - \bar{T}_{\bullet})^2$$

We compare the value of Q to a chi-square distribution with $k - 1$ degrees of freedom. When the value of Q exceeds the $100(1-\alpha)$ percentile of the central chi-square distribution with $k - 1$ degrees of freedom, we reject the null hypothesis, and report that our variance component is different from zero.

5.3.1   Power of the Test of the Significance of the Variance Component in Random Effects Models

As I discuss above, when the null hypothesis is true, i.e., when $H_0 : \tau^2 = 0$, the statistic Q has the chi-square distribution with $k - 1$ degrees of freedom. However, when the null hypothesis is false, then Q has a complex distribution under the random effects model. This distribution is a weighted combination of chi-square distributions and must be approximated. As given in Hedges and Vevea (1998) and in Hedges and Pigott (2001), one approximation, derived by Satterthwaite (1946), uses a gamma distribution with mean and variance equal to functions of the mean and variance of the distribution of Q, this weighted combination of chi-square distributions. The mean of the Q distribution is given by μ_Q and can be written as

$$\mu_Q = c\,\tau^2 + (k - 1) \tag{5.5}$$

where c is given in (2.5), and τ^2 is the value of the variance component that we are testing as significantly different from 0. The variance of Q, σ_Q^2, is computed using the fixed effects weights and is given by

$$\sigma_Q^2 = 2(k - 1) + 4\left(\sum w_i - \frac{\sum w_i^2}{\sum w_i}\right)\tau^2$$
$$+ 2\left(\sum w_i^2 - 2\frac{\sum w_i^3}{\sum w_i} + \frac{\left(\sum w_i^2\right)^2}{\left(\sum w_i\right)^2}\right)\tau^4 \tag{5.6}$$

We can then use a central chi-square distribution with non-integer degrees of freedom to approximate the gamma distribution with mean given in (5.5), and variance in (5.6). To compute the correct approximation, we need to compute two more quantities based on the mean and variance of given in (5.5 and 5.6). These quantities are r, given by

$$r = \frac{\sigma_Q^2}{2\mu_Q} \tag{5.7}$$

and s, given by

$$s = \frac{2(\mu_Q)^2}{\sigma_Q^2} \,. \tag{5.8}$$

Then the power of the test when $\tau^2 = 0$ is given by

$$F(c_\alpha/r \,|\, s; 0) \tag{5.9}$$

where $F\,(c_\alpha/r \,|\, s; 0)$ is the cumulative distribution function of the central chi-square with s degrees of freedom (a non-integer value), and c_α is the $100(1 - \alpha)$ percentile point of the chi-square distribution with $(k - 1)$ degrees of freedom. Tabled values of the chi-square and functions that compute the chi-square cumulative distribution traditionally give the area under the curve that is larger than c_α, the critical value. This area larger than the critical value is the power of the test. The Appendix gives options for computing the distribution of a central chi-square distribution with non-integer degrees of freedom.

5.3.2   Choosing Values for the Parameters Needed to Compute the Variance Component in Random Effects Models

As seen in Sect. 5.2.1, the test of the variance component requires the researcher to pose a value for the variance component that the reviewer wishes to test as significantly different from 0. Section 4.4.2 presents a method for suggesting values of the variance component, τ^2, using conventions suggested by Higgins and Thompson (2002). Recall that we can use conventions based on values of I^2 corresponding to 25% for low heterogeneity, 50% for moderate heterogeneity, and 100% for a large degree of heterogeneity. Given the relationship in (4.10), where

$$I^2 = \frac{\tau^2}{\tau^2 + v}$$

and v is our common value for the within-study sampling variance of the effect size, we can pose values for τ^2 equal to $(1/3)v$, v, and $3v$, as our values for low, moderate, and large degrees of heterogeneity, respectively. As seen in (5.6), we will also need a value for the common fixed effects weights, $w = 1/v$. We computed the common value of v by assuming that each study in the meta-analysis has the same sample size, and in the case of the standardized mean difference, that each study has the same effect size. This simplifying assumption allows us to compute a common value for the sampling variance, v, of the effect size of choice. Example 4.3.4 illustrates how we pose a value for v.

5.3.3 Example: Computing Power for Values of τ^2, the Variance Component

In Chap. 4, we used an example of studies on the effectiveness of coaching programs to increase the SAT-Math and SAT-Verbal scores of high school students. Let us say that we are interested in a standardized mean difference of 0.2, and assume that a typical study has a total sample size of 40, with 20 participants each in the experimental and control groups. We obtain a value of the common effect size of v below as

$$v = \frac{20 + 20}{20 * 20} + \frac{(0.2)^2}{2(20 + 20)} = 0.10$$

Given a value of v, we can compute the power for values of τ^2. For example, with a small degree of heterogeneity, we would compute $\tau^2 = (1/3)v = 0.033$, and with a large degree of heterogeneity, we would have $\tau^2 = 3(0.10) = 0.30$. With the number of studies equal to $k = 10$, we can compute c, given in (2.5), and equal to

$$c = \sum_{i=1}^{10} w - \frac{\sum_{i=1}^{10} w^2}{\sum_{i=1}^{10} w} = 10w - \frac{10w^2}{10w} = \frac{10}{0.10} - \frac{(1/0.10)^2}{(1/0.10)}$$

$$= 100 - 10 = 90.$$

Let us begin with the computations assuming a small degree of heterogeneity. With $k = 10$ studies, and $c = 90$, we compute a value of $\mu_Q = 90(0.033) + (10-1) = 2.97 + 9 = 11.97$. Computing σ_Q^2 is more computationally intense as seen in (5.6), and is given by

$$\sigma_Q^2 = 2(10 - 1) + 4\left(\sum 1/0.10 - \frac{\sum (1/0.10)^2}{\sum 1/0.10}\right)(0.033)^2$$

$$+ 2\left(\sum (1/0.10)^2 - 2\frac{\sum (1/0.10)^3}{\sum 1/0.10} + \frac{\left(\sum (1/0.10)^2\right)^2}{\left(\sum 1/0.10\right)^2}\right)(0.033)^4$$

$$= 18 + 0.39 + 2[1000 - 200 + 100](0.001)^2$$

$$= 18.39 + 0.002 = 18.392 \tag{5.10}$$

Using (5.7), $r = 18.392/(2*11.97) = 0.77$, and $s = 2(11.97)^2/18.392 = 15.58$. With values of r, and s, we can compute the power of the test of the variance component when $\tau^2 = 0.033$, a small degree of heterogeneity in this case. Our critical value for the central chi-square distribution with degrees of freedom,

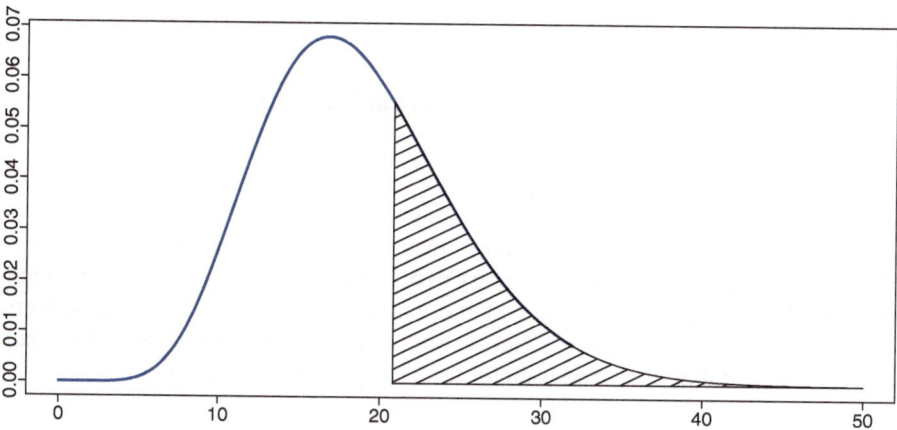

Fig. 5.2 Chi-square distribution with 18.8 degrees of freedom

$k - 1 = 9$, and $\alpha = 0.05$ is $c_\alpha = 16.92$. The power of the test of the variance component using (5.9), and $\alpha = 0.05$ is $F(c_\alpha/r \mid s; 0) = F(16.92/0.77 \mid 15.58; 0) = 0.13$. We have little power to find $\tau^2 = 0.033$ significantly different from zero in this example.

If we assume a large degree of heterogeneity, in this case, that $\tau^2 = 0.3$, we have $\mu_Q = 90(0.3) + (10 - 1) = 27 + 9 = 36$. For σ_Q^2, we obtain

$$\sigma_Q^2 = 2(10 - 1) + 4\left(\sum 1/0.10 - \frac{\sum (1/0.10)^2}{\sum 1/0.10}\right)(0.3)^2$$

$$+ 2\left(\sum (1/0.10)^2 - 2\frac{\sum (1/0.10)^3}{\sum 1/0.10} + \frac{\left(\sum (1/0.10)^2\right)^2}{\left(\sum 1/0.10\right)^2}\right)(0.3)^4$$

$$= 18 + 32.4 + 2[1000 - 200 + 100](0.09)^2$$
$$= 50.4 + 14.58 = 64.98 \tag{5.11}$$

Using (5.7), $r = 64.98/(2*36) = 0.90$, and $s = 2(36)^2/64.98 = 20.81$. With values of r, and s, we can compute the power of the test of the variance component when $\tau^2 = 0.3$, a large degree of heterogeneity in this case. As above, our critical value is $c_\alpha = 16.92$. The power of the test of the variance component using (5.9), is $F(c_\alpha/r \mid s; 0) = F(16.92/0.90 \mid 20.81; 0) = 0.59$. We have more power to find $\tau^2 = 0.3$ significantly different from zero than in the prior example, though we still do not reach the level of power $= 0.8$. Figure 5.2 shows a chi-square distribution with 18.8 degrees of freedom. The area to the right of $x = 20.81$ is the power in this example. See the Appendix for computing options to obtain the values of a chi-square with non-integer degrees of freedom.

Appendix

Computing Power for the Tests of Homogeneity and the Variance Component

The examples in this chapter require the computation of the area under the non-central chi-square that is either less than or greater than a given critical value, c_α. Below I give the functions and or commands necessary to obtain the power for the test of homogeneity in fixed effects, and the test of the variance component in random effects. The program Exel only provides the values of the cumulative central chi-square distribution, and thus cannot be used for the examples in this chapter.

SPSS

Using the Compute menu in SPSS, the function IDF.CHISQ(p, df) provides the critical value c_α for a central chi-square distribution where p equals the area in the left tail, and df equals the degrees of freedom. The following command returns the critical value for the example in Sect. 5.2.3.

$$critval = IDF.CHISQ(0.95, 9)$$

SPSS gives the critical value for the command above as 16.919. The function given by NCDF.CHISQ($quant$, df, nc) gives the area less than or equal to quant $= c_\alpha$ for a non-central chi-square distribution with df degrees of freedom and non-centrality parameter equal to nc. To compute the power for Example 5.2.3,

$$power = 1 - NCDF.CHISQ(16.92, 9, 4)$$

returns the value 0.225.

The example in Sect. 5.3.3 uses the central chi-square distribution with non-integer degrees of freedom. The command in SPSS

$$power = 1 - CDF.CHISQ(21.974, 15.58)$$

returns the value 0.128.

SAS

In SAS, the function CINV(p, df) provides the critical value, c_α, of the cumulative central chi-square distribution function with df degrees of freedom that is less than c_α. The function CDF('CHISQUARE', x, df, ncp) provides the area in the left tail that is less than or equal to x in a non-central chi-square distribution with df degrees of freedom and non-centrality parameter equal to ncp. To compute the power for the example in Sect. 5.2.3, we first obtain the critical value for the central chi-square distribution with 9° of freedom at the $\alpha = 0.05$ level as below

$$critval = cinv(0.95, 9).$$

The command above returns a value of 16.9190. The power of the first test in Sect. 5.2.3 is computed in SAS as

$$power = 1 - cdf('chisquare', 16.92, 9, 4)$$

The above function results in a value of 0.22536. The example in Sect. 5.3.3 requires the central chi-square distribution with non-integer degrees of freedom. Power for the first example in Sect. 5.2.3 is computed as

$$power = 1 - cdf('chisquare', 21.974, 15.58, 0)$$

This command results in a value of 0.12834.

R

In R, the function qchisq(p, df, ncp, lower.tail = TRUE, log.p = FALSE) gives the critical value for the cumulative chi-square distribution with df degrees of freedom for the area equal to p. The option lower.tail = TRUE indicates that we want p to equal the area in the lower tail. The option log.p = FALSE indicates that we do not want the value of p in log units. The command below computes the critical value for the 95% point of the chi-square distribution with 9 degrees of freedom and non-centrality parameter = 0. When the non-centrality parameter is equal to 0, the distribution is a central chi-square distribution.

```
> qchisq(0.95, 9, ncp = 0, lower.tail = TRUE, log.p = FALSE)
[1]16.91898
```

To obtain the power for the second computation in Example 5.2.3, we use the command

```
> pchisq(16.919, 9, ncp = 9, lower.tail = FALSE, log.p = FALSE)
[1]0.5104429
```

Note that in this example, we are asking for the area in the upper tail as indicated by the option lower.tail = FALSE. The first example in Sect. 5.3.3 is computed as

> pchisq(21.97, 15.58, ncp = 0, lower.tail = FALSE, log.p = FALSE)
[1] 0.1284637

As in the prior example, we want the upper tail, so lower.tail = FALSE.

References

Hedges, L.V., and T.D. Pigott. 2001. Power analysis in meta-analysis. *Psychological Methods* 6: 203–217.
Hedges, L.V., and J.L. Vevea. 1998. Fixed- and random-effects models in meta-analysis. *Psychological Methods* 3(4): 486–504.
Higgins, J.P.T., and S.G. Thompson. 2002. Quantifying heterogeneity in a meta-analysis. *Statistics in Medicine* 21: 1539–1558.
Satterthwaite, F.E. 1946. An approximate distribution of estimates of variance components. *Biometrics Bulletin* 2: 110–114.

Chapter 6
Power Analysis for Categorical Moderator Models of Effect Size

Abstract This chapter provides methods for computing power with moderator models of effect size. The models discussed are analogues to one-way ANOVA models for effect sizes. Examples are provided for both fixed and random effects categorical moderator models. The power for meta-regression models requires knowledge of the values of the predictors for each study in the model, and is not provided here.

6.1 Background

The prior two chapters outlined the procedures for computing power to test the mean effect size and homogeneity in both fixed and random effects models. As discussed in Chap. 3, reviewers plan many syntheses to test theories about why effect sizes differ among studies. These theories are formally examined in the form of moderator analyses in meta-analysis. Given the variability we expect among studies especially in the social sciences, moderator analyses provide important information about, for example, how the effects of an intervention vary in different contexts and with different participants. Computing the power of moderator analyses is critical when planning a meta-analysis, since examining potential reasons for variance among effect sizes is an important focus of a systematic review. This chapter will provide the computations for power for categorical models of effect sizes that are analogous to one-way ANOVA. These models can be computed under both the fixed or random effects assumptions, and I discuss the power calculations under both assumptions.

As we have seen in the prior chapters, we face challenges in posing important values for the parameters needed in the power computations. For categorical models, we will need to arrive at values for the differences among the means of groups of effect sizes, and for the degree of heterogeneity within groups. When using random effects models, we also need to provide plausible values for the variance component, τ^2. Both fixed effects and random effects models will be discussed as well as strategies for providing values for the power parameters.

T.D. Pigott, *Advances in Meta-Analysis*, Statistics for Social and Behavioral Sciences,
DOI 10.1007/978-1-4614-2278-5_6, © Springer Science+Business Media, LLC 2012

6.2   Categorical Models of Effect Size: Fixed Effects One-Way ANOVA Models

The simplest moderator models for effect size are analogues to one-way ANOVA models where the researcher is interested in comparing the mean effect sizes across groups of studies. The groups of studies are formed from a small number of levels of a categorical factor. For example, one moderator included in many syntheses is whether the study used random assignment to place individuals into experimental groups. Computation of categorical models with a single factor in meta-analysis proceeds in the same manner as one-way ANOVA. In meta-analysis, instead of computing the sums of squares, we compute between-group and within-group homogeneity tests. Below I outline the procedures for computing the statistics for the one-way fixed effects ANOVA model in meta-analysis, followed by a discussion of how to compute power.

6.2.1   Tests in a Fixed Effects One-Way ANOVA Model

As discussed in Chap. 2, we refer to our set of effect sizes using T_i, $i = 1, 2, \ldots, k$, where k is the total number of studies. Now let us assume that our k studies fall into p groups as defined by a moderator variable. For example, the moderator variable might be the grade level where the intervention takes place, say elementary or high school. We assume that p is a small number of categories. We can then designate m_i as the number of studies in the ith group, so that k is the total number of studies, $k = m_1 + m_2 + \ldots + m_p$.

Our interest in a one-way ANOVA effect size model is to examine whether the means for the groups are equal. We will also want to know if the effect sizes within each group are homogeneous. We will compute two values of the Q statistic, Q_B, the test of between-group homogeneity, and Q_W, the test of within-group homogeneity. The Q_B is analogous to the F-test between groups, and is an omnibus test of whether all group means are equal. The Q_W is an overall test of within-group homogeneity, and is equal to the sum of the homogeneity tests within each group. Thus, $Q_W = Q_{W_1} + Q_{W_2} + \ldots + Q_{Wp}$ where $Q_{W_i}, i = 1, 2, \ldots, p$ are the tests of homogeneity within each group. In addition, the two tests of homogeneity, like the F-tests in ANOVA, sum to the overall homogeneity test across all effect sizes, or, $Q_T = Q_B + Q_W$. The power computations for these two tests of homogeneity are given below.

6.2.2   Power of the Test of Between-Group Homogeneity, Q_B, in Fixed Effects Models

The first concern in the ANOVA model is whether the group mean effect sizes are equal. When there are p group means, the omnibus test of the null hypothesis that the group mean effect sizes are equal is given by

$$H_0 \; : \; \bar{\theta}_{1\bullet} = \bar{\theta}_{2\bullet} = ... = \bar{\theta}_{p\bullet} \tag{6.1}$$

To test this hypothesis, we compute the between-groups homogeneity test given by

$$Q_B = \sum_{i=1}^{p} w_{i\bullet} \, (\bar{T}_{i\bullet} - \bar{T}_{\bullet\bullet})^2 \tag{6.2}$$

where $w_{i\bullet}$ is the sum of the weights in the ith group, $\bar{T}_{i\bullet}$ is the mean effect size in the ith group, and $\bar{T}_{\bullet\bullet}$ is the overall mean effect size. These quantities are more formally given as

$$w_{i\bullet} = \sum_{j=1}^{m_i} w_{ij} \, , \;\; \text{where } i = 1, ..., p, \text{ and } j = 1, ..., m_i$$

$$\bar{T}_{i\bullet} = \frac{\sum\limits_{j=1}^{m_i} w_{ij} T_{ij}}{\sum\limits_{j=1}^{m_i} w_{ij}}$$

$$\bar{T}_{\bullet\bullet} = \frac{\sum\limits_{i=1}^{p} \sum\limits_{j=1}^{m_i} w_{ij} T_{ij}}{\sum\limits_{i=1}^{p} \sum\limits_{j=1}^{m_i} w_{ij}} \tag{6.3}$$

When the null hypothesis in (6.1) is true, that is when all the group means are equal, Q_B has the chi-square distribution with $(p - 1)$ degrees of freedom. When $Q_B > c_\alpha$ where c_α is the $100(1 - \alpha)$ percentile point of the chi-square distribution with $(p - 1)$ degrees of freedom, we reject the null hypothesis. When the null hypothesis is false, that is when at least one of the means differs from the other group means, Q_B has a non-central chi-square distribution with $(k - 1)$ degrees of freedom and non-centrality parameter λ_B given by

$$\lambda_B = \sum_{i=1}^{p} w_{i\bullet} \, (\bar{\theta}_{i\bullet} - \bar{\theta}_{\bullet\bullet})^2. \tag{6.4}$$

The power of the test of Q_B is

$$1 - F(c_\alpha \,|\, p - 1; \lambda_B) \tag{6.5}$$

where $F(c_\alpha \,|\, p\text{-}1; \lambda_B)$ is the cumulative distribution of the non-central chi-square with $(p - 1)$ degrees of freedom and non-centrality parameter, λ_B.

Table 6.1 Steps for computing power in fixed effects one-way ANOVA model

1. Establish a critical value for statistical significance, c_α
2. Decide on the magnitude of the difference between the group means in the effect size metric used. For example, with standardized mean differences, decide on the number of standard deviations that are substantively important. For correlations, decide on the difference in correlations that is important
3. Assign values to the group means, $\theta_1,\ldots,\theta_p$ corresponding to the differences between group means in (2)
4. Estimate the number of studies within each group, $m_1,\ldots,m_p$
5. Compute w_{ij}, the common value of the weights for each effect size, given "typical" values for the within-study sample sizes

6.2.3 Choosing Parameters for the Power of Q_B in Fixed Effects Models

As in prior chapters, our challenge is to pose important values for the parameters in the power computations. Our main interest in the test of between-group homogeneity is the difference among the group effect sizes. Thus, we can pose a substantively important difference that we would like to test among the group means. If we, for example, want to know if the mean effect size for studies that use random assignment is at least 0.5 standard deviations smaller than the mean effect size for quasi-experiments, we could assume that the group mean effect size for the randomized studies is 0.0, and that for the quasi-experimental studies is 0.5. Thus, to test the between-group homogeneity, we can pose values for the mean effect sizes in each group, or at least the difference we would like to test between groups. We then need to suggest the number of effect sizes we will have in each group, and compute values for w_{ij}, the weights for the individual effect sizes. In prior chapters, we have made the simplifying assumption that all studies have the same sample sizes to obtain a common value of the w_{ij}, by assuming a "typical" within-study sample size. Table 6.1 summarizes the values needed to compute power for a fixed effects categorical moderator model.

6.2.4 Example: Power of the Test of Between-Group Homogeneity in Fixed Effects Models

One of the data sets used in the book is based on Sirin (2005), a meta-analysis of the correlations between measures of socio-economic status and academic achievement. The studies included in this meta-analysis used samples of students at several grade levels. Let us say that we want to make sure our meta-analysis will have enough power to detect differences in the mean effect size for studies that use students at three different grade levels: elementary (K-3), middle (4–8), and high school (9–12). We may be interested in whether the difference in the mean correlation between elementary and middle school is 0.1, and between elementary and high school is 0.5 (implying a difference of 0.4 for middle versus high school). We have three groups, $p = 3$, and we can assume that our mean effect sizes are $\bar{\theta}_1 = 0.0$, $\bar{\theta}_2 = 0.1$, and $\bar{\theta}_3 = 0.5$. In terms of Fisher's z-transformations, these means would be equal to $\bar{\theta}_{z1} = 0.0$, $\bar{\theta}_{z2} = 0.10$, and $\bar{\theta}_{z3} = 0.55$, respectively. Let us also assume that we have 5 studies per group, and our common within-group sample size is $n = 15$. To compute the non-centrality parameter, λ_B, we first need our common value of w_{ij}. For Fisher's z, $w_{ij} = 1/v_{ij} = n_{ij} - 3$ (v_{ij} for correlations is given in 2.15). Thus, we can compute

$$w_{i\bullet} = \sum_{j=1}^{5} w_{ij} = \sum_{j=1}^{5} (15 - 3) = \sum_{j=1}^{5} 12 = 60$$

as given in (6.3). The overall mean effect size is $\bar{\theta}_{i\bullet} = (0.0 + 0.10 + 0.55)/3 = 0.22$, since all effect sizes have the same weight, and we have equal numbers of studies within each group. (With different weights for each effect size and different numbers of effect sizes within groups, we would need to use a weighted mean as given in (6.3). We can compute the non-centrality parameter as

$$\lambda_B = \sum_{i=1}^{3} w_{i\bullet} (\bar{\theta}_{i\bullet} - \bar{\theta}_{\bullet\bullet})^2$$
$$= 60(0.0 - 0.22)^2 + 60(0.10 - 0.22)^2 + 60(0.55 - 0.22)^2$$
$$= 60(-0.22)^2 + 60(-0.12)^2 + 60(0.33)^2$$
$$= 10.302$$

The central chi-square distribution with $p - 1 = 3-1 = 2$ degrees of freedom has a critical value equal to 5.99 with $\alpha = 0.05$. The power of the omnibus test that $H_0 : \bar{\theta}_{z1} = \bar{\theta}_{z2} = \bar{\theta}_{z3}$ is given by $1 - F(5.99 \mid 2; 10.302) = 1 - 0.17 = 0.83$. The Appendix in Chap. 5 provides options for computing values of the non-central chi-square distribution.

6.2.5 Power of the Test of Within-Group Homogeneity, Q_W, in Fixed Effects Models

If a reviewer finds that the effect size means do differ between groups, then a second question centers on whether the effect sizes within those groups are homogeneous. The rationale for this question is similar to that for the overall test of homogeneity. While the mean effect sizes may differ among groups, we also need to know if the effect sizes within the group are estimating a common mean effect size. Hedges and Pigott (2004) refer to the test of within-group homogeneity as a test of the goodness of fit of the fixed effects model. In other words, the ANOVA model proposed fits the data well if the effect sizes within groups are homogeneous, i.e., that the effect sizes within each group estimate a common mean value. The overall test of within-group homogeneity is the omnibus test that the effect sizes within each group estimate a common mean. We can write the null hypothesis as

$$H_0 \; : \; \theta_{ij} = \bar{\theta}_{i\bullet}, \;\; i = 1, ..., p; \, j = 1, ..., m_i \tag{6.6}$$

where the alternative hypothesis is that at least one of the effect sizes in group i differs from the group mean, and there are m_i effect sizes within the ith group. The test for overall within-group homogeneity is

$$Q_W = \sum_{i=1}^{p} \sum_{j=1}^{m_i} w_{ij} \left(T_{ij} - \bar{T}_{i\bullet}\right)^2 = \sum_{i=1}^{p} Q_{W_i}. \tag{6.7}$$

Note that this sum can also be written as the sum of the within-group homogeneity statistics, Q_{W_i}. When every group in the model is homogeneous, Q_W has the central chi-square distribution with $(k - p)$ degrees of freedom, where k is the number of effect sizes and p is the number of groups. We reject the null hypothesis when, $Q_W > c_\alpha$ where c_α is the $100(1 - \alpha)$ percentile point of the chi-square distribution with $(k - p)$ degrees of freedom. Rejecting the null hypothesis indicates that at least one group is heterogeneous, i.e., that at least one effect size differs significantly from its group mean.

When the null hypothesis is false, the statistic Q_W has a non-central chi-square distribution with $(k - p)$ degrees of freedom, and non-centrality parameter λ_W given by

$$\lambda_W = \sum_{i=1}^{p} \sum_{j=1}^{m_i} w_{ij} \left(\theta_{ij} - \bar{\theta}_{i\bullet}\right)^2. \tag{6.8}$$

The power of the test of Q_W is given by

$$1 - F(c_\alpha \,|\, k - p; \, \lambda_W), \tag{6.9}$$

where $F(c_\alpha \mid k - p; \lambda_W)$ is the cumulative distribution of the non-central chi-square with $(k - p)$ degrees of freedom and noncentrality parameter λ_W, evaluated at c_α, the desired critical value of the central χ^2.

6.2.6   Choosing Parameters for the Test of Q_W in Fixed Effects Models

The difficulty in computing the power of within-group heterogeneity is in posing a substantively important value of heterogeneity. As we did in Chap. 5, we could decide on the amount of heterogeneity we would want to detect within groups based on the standard error of the group mean effect size. For example, we might want to determine the power to detect heterogeneity if one of the groups had effect sizes that differed from the mean by 3 standard errors of the mean. Below I illustrate an example of this strategy.

6.2.7   Example: Power of the Test of Within-Group Homogeneity in Fixed Effects Models

Let us continue with the example in Sect. 6.2.4. In that example, we have three groups, $p = 3$, corresponding to studies with elementary, middle and high school students, respectively. The mean Fisher z-transformations for each group are given by $\bar{\theta}_{z1} = 0.0$, $\bar{\theta}_{z2} = 0.10$, and $\bar{\theta}_{z3} = 0.55$, respectively. Let us also assume that we have 5 studies per group (for a total $k = 15$), and our common within-group sample size is $n = 15$. Thus, we have a common effect size variance of $v = 1/(n-3) = 1/(15-3) = 1/12$, with a common weight equal to $w = 1/(1/12) = 12$. As we did in Chap. 5, we can decide on how much variation we expect within the groups for our power computations. Given our within-group sample sizes, we can compute the variance (and standard error) of our mean effect sizes as

$$v_{i\bullet} = \frac{1}{\sum\limits_{j=1}^{5} w} = \frac{1}{\sum\limits_{j=1}^{5} 12} = \frac{1}{60} = 0.017$$

$$\sqrt{v_{i\bullet}} = \sqrt{0.017} = 0.13$$

Let us say that the effect sizes using elementary school samples and those from middle school samples differ by one standard deviation on average from their group mean, while the effect sizes using high school samples differ by four standard deviations on average from its group mean. We can compute the non-centrality parameter, λ_W, from (6.8) as

$$\lambda_W = \sum_{i=1}^{p} \sum_{j=1}^{m_i} w_{ij} \left(\theta_{ij} - \bar{\theta}_{i\bullet} \right)^2$$

$$= \sum_{j=1}^{5} 12(0.13)^2 + \sum_{j=1}^{5} 12(0.13)^2 + \sum_{j=1}^{5} 12(4*0.13)^2$$

$$= 5 * 12(0.017) + 5 * 12(0.017) + 5 * 12(0.52)^2$$

$$= 60(0.017) + 60(0.017) + 60(0.27)$$

$$= 1.02 + 1.02 + 16.20$$

$$= 18.24$$

In the computation above, we replace the difference between the individual effect sizes and their group mean, $(\theta_{ij} - \bar{\theta}_{i\bullet})^2$, by the proposed number of standard deviations, $\sqrt{v_{i\bullet}}$ among the means. To compute power, we need the $c_{0.05}$ critical value of a central chi-square with $k - p = 15 - 3 = 12$ degrees of freedom, which is equal to 21.03. Equation 6.9 gives the power as $1 - F(c_\alpha | k - p; \lambda_W) = 1 - F(21.03 | 12; 18.24) = 1 - 0.18 = 0.82$. In this example, we have adequate power to find that at least one of our groups of effect sizes is not homogeneous. The Appendix in Chap. 5 provides the program code needed to compute values for the non-central chi-square distribution.

6.3   Categorical Models of Effect Size: Random Effects One-Way ANOVA Models

As discussed in Chap. 3, a random effects model for effect sizes assumes that each study's effect size is a random draw from a population of effect sizes. Thus, each effect size differs from the overall mean effect size due to the underlying variance of the population, designated as τ^2, and due to within-study sampling variance, v_i. Our goal in the random effects analysis is to first compute the variance component, τ^2, and then to use the variance component to compute the random effects weighted mean effect size.

When we are interested in estimating a random effects categorical model for moderators, we will conduct a similar analysis for each group defined by our categorical factor. As indicated in Chap. 3, we will make the assumption that there is a common variance component, τ^2, across studies, regardless of the study's value for the categorical factor. Given that we will assume the same variance component across studies, we will only focus on the test of between-group heterogeneity. The test of the significance of the variance component given in Chap. 5 for the random effects model would apply to the case where we are assuming a common variance component across studies.

As in Sect. 6.2.1, our k studies fall into p groups as defined by a moderator variable where, $k = m_1 + m_2 + ... + m_p$. The effect size in the jth study in the ith group is then designated by T_{ij}^* with variance of v_{ij}^*. The variance of each effect size contains two components, one due to sampling error in an individual study, denoted by v_{ij} and one due to the variance component, τ^2. The random effects variance for the effect size T_{ij}^* can be written as $v_{ij}^* = v_{ij} + \tau^2$.

6.3.1   Power of Test of Between-Group Homogeneity in the Random Effects Model

Similar to the fixed effects case, our test for between-group mean differences is an omnibus test. The null hypothesis for the test that the random effects group means differ can be written as

$$H_0 : \theta_{1\bullet}^* = \theta_{2\bullet}^* = ... = \theta_{p\bullet}^* \qquad (6.10)$$

where the $\theta_{i\bullet}^*$ the random effects means for the $i = 1,..., p$ groups. We test this hypothesis by computing the between-groups random effects homogeneity test given by

$$Q_B^* = \sum_{i=1}^{p} w_{i\bullet}^* \, (\bar{T}_{i\bullet}^* - \bar{T}_{\bullet\bullet}^*)^2 . \qquad (6.11)$$

where $w_{i\bullet}^*$ is the sum of the weights in the ith group, $\bar{T}_{i\bullet}^*$ is the mean effect size in the ith group, and $\bar{T}_{\bullet\bullet}^*$ is the overall mean effect size, all in the random effects model. These quantities are more formally given as

$$w_{i\bullet}^* = \sum_{j=1}^{m_1} w_{ij}^*, \quad \text{where } i = 1, ..., p, \text{ and } j = 1, ..., m_i$$

$$\bar{T}_{i\bullet}^* = \frac{\sum_{j=1}^{m_i} w_{ij}^* T_{ij}^*}{\sum_{j=1}^{m_i} w_{ij}^*}$$

$$\bar{T}_{\bullet\bullet}^* = \frac{\sum_{i=1}^{p} \sum_{j=1}^{m_i} w_{ij}^* T_{ij}^*}{\sum_{i=1}^{p} \sum_{j=1}^{m_i} w_{ij}^*}$$

When the null hypothesis is true, i.e., when all the means are equal, Q_B^* is distributed as a central chi-square distribution with $(p - 1)$ degrees of freedom. When the value of Q_B^* exceeds the critical value c_α which is the $100(1 - \alpha)$ percentile point of the central chi-square distribution with $(p - 1)$ degrees of freedom, we assume that at least one of the random effects group means is significantly different from the rest of the means. In the case where we reject the null hypothesis, Q_B^* has a non-central chi-square distribution with $(p - 1)$ degrees of freedom, and non-centrality parameter λ_B^* given by

$$\lambda_B^* = \sum_{i=1}^{p} w_{i\bullet}^* \, (\theta_{i\bullet}^* - \theta_{\bullet\bullet}^*)^2. \qquad (6.12)$$

The power of the test for the between-group mean difference in a random effects model is given as

$$1 - F(c_\alpha \,|\, p - 1; \, \lambda_B^*) \qquad (6.13)$$

where $F(c_\alpha \,|\, p - 1; \, \lambda_B^*)$ is the cumulative distribution function at c_α of the non-central chi-square with $(p - 1)$ degrees of freedom and non-centrality parameter λ_B^*.

Table 6.2 Steps for computing power in random effect categorical moderator model

1. Establish a critical value for statistical significance, c_α
2. Decide on the magnitude of the difference between the group means in the effect size metric used. For example, with standardized mean differences, decide on the number of standard deviations that are substantively important. For correlations, decide on the difference in correlations that is important
3. Assign values to the group means, $\theta_1, \ldots, \theta_p$ corresponding to the differences between group means in (2)
4. Estimate the number of studies within each group, $m_1, \ldots, m_p$
5. Compute v, the common value of the sampling variance for each effect size, given "typical" values for the within-study sample sizes
6. Compute τ^2 for levels of heterogeneity based on v. Large degree of heterogeneity is $\tau^2 = 3v$, moderate level is $\tau^2 = v$, and a small degree is $\tau^2 = (1/3)v$
7. Compute the values for the common weight, $w = 1/(v + \tau^2)$

6.3.2 Choosing Parameters for the Test of Between-Group Homogeneity in Random Effects Models

To compute the power of the test of between-group mean differences, we will need to decide on the size of the difference between the group means that is of substantive importance. In random effects models, we will also need to decide how much variation we have between studies, i.e., we need a value for τ^2. We can use the convention based on Higgins and Thompson (2002). Recall in Chap. 4 that we chose a value of τ^2 based on the value of v, the "typical" within-study sampling variance of effect sizes. A large degree of heterogeneity, and thus a large value of τ^2, was assumed to be $\tau^2 = 3v$, which corresponds to an I^2 value of .75. A moderate degree of heterogeneity is $\tau^2 = v$, corresponding to an I^2 value of .5, and a small degree of heterogeneity is $\tau^2 = (1/3)v$, corresponding to an I^2 value of .25. We can amend the steps in Table 6.1 for the power for differences of random effects means as seen in Table 6.2.

Thus, to compute the power of the test of the between-group random effects means, we need to pose values for the number of studies within each group, the within-study sample size, the degree of heterogeneity we expect, and the magnitude of the differences between the means that is substantively important.

6.3.3 Example: Power of the Test of Between-Group Homogeneity in Random Effects Models

Let us return to the example in Sect. 6.2.4. Recall that we have three groups, $p = 3$, and we can assume that our mean effect sizes are $\bar{\theta}_1 = 0.0$, $\bar{\theta}_2 = 0.1$, and $\bar{\theta}_3 = 0.5$. In terms of Fisher's z-transformations, these means would be equal to $\bar{\theta}_{z1} = 0.0$, $\bar{\theta}_{z2} = 0.10$, and $\bar{\theta}_{z3} = 0.55$, respectively. Let us also assume that we have 5 studies per group (with $k = 15$), and our common within-group sample size is $n = 15$. To compute the non-centrality parameter λ_B^*, we first need to posit a value of the common value of the variance component, τ^2, across studies, based on our common value of v. For Fisher's z-transformation, the common value of the within-group sampling variance in this example is $v = 1/(n - 3) = 1/12 = 0.083$. For a large degree of heterogeneity, the variance component would equal $\tau^2 = 3$ $(0.083) = 0.25$, a moderate degree of heterogeneity, $\tau^2 = 0.083$, and a small degree of heterogeneity, $\tau^2 = (1/3)0.083 = 0.028$. Thus, our value for the random effects variance, v^*, is given as $v^* = 0.083 + \tau^2$. Assuming a small degree of heterogeneity, the value of the weight for the random effects means, $w_{i\bullet}^*$, is given as

$$w_{i\bullet}^* = \frac{1}{\sum_{i=1}^{5} \frac{1}{v + \tau^2}} = \frac{1}{\sum_{i=1}^{5} \frac{1}{0.083 + 0.028}} = \frac{1}{5 * 9.01} = 0.022$$

Given our value of $w_{i\bullet}^*$, we can compute the non-centrality parameter λ_B^* as

$$\lambda_B^* = \sum_{i=1}^{3} w_{i\bullet}^* (\theta_i^* - \theta_\bullet^*)^2$$
$$= 0.022(0 - 0.22)^2 + 0.022(0.1 - 0.22)^2 + 0.022(0.55 - 0.22)^2$$
$$= 0.022(0.048) + 0.022(0.014) + 0.022(0.11)$$
$$= 0.004$$

The power of the test of the between-group differences is given as $1 - F \times (c_\alpha | p - 1; \lambda_B^*)$ in (6.13). With $p - 1 = 2$ degrees of freedom, the $c_{0.05}$ critical value of a central chi-square distribution is 5.99. Thus, the power in this example with a small degree of heterogeneity is $1 - F(5.99 | 2; 0.004) = 1 - 0.95 = 0.05$.

We can also compute the power with a large degree of heterogeneity as a comparison. With a large degree of heterogeneity, we have a value for the random effects variance, v^*, as $v^* = 0.083 + 0.25 = 0.33$. Assuming a large degree of heterogeneity, the value of the weight for the random effects means, $w_{i\bullet}^*$, is given as

$$w_{i\bullet}^* = \frac{1}{\sum_{i=1}^{5} \frac{1}{v + \tau^2}} = \frac{1}{\sum_{i=1}^{5} \frac{1}{0.33}} = \frac{1}{5 * 3.033} = \frac{1}{15.01} = 0.067$$

Given our value of $w_{i\bullet}^*$, we can compute the non-centrality parameter λ_B^* as

$$\lambda_B^* = \sum_{i=1}^{3} w_{i\bullet}^* (\theta_i^* - \theta_\bullet^*)^2$$
$$= 0.067(0 - 0.22)^2 + 0.067(0.1 - 0.22)^2 + 0.067(0.55 - 0.22)^2$$
$$= 0.067(0.048) + 0.067(0.014) + 0.067(0.11)$$
$$= 0.012$$

With $p - 1 = 2$ degrees of freedom, the $c_{0.05}$ critical value of a central chi-square distribution is 5.99. Thus, the power in this example with a large degree of heterogeneity is $1 - F(5.99 \mid 2; 0.012) = 1 - 0.95 = 0.05$. In the random effects model, we have little power to detect a difference among the mean effect sizes.

6.4   Linear Models of Effect Size (Meta-regression)

When a reviewer wishes to test a model with a number of moderators, including continuous predictors, the relatively simple ANOVA models discussed above have limited utility. Large-scale reviews often include multiple moderators that reviewers would like to test simultaneously in a linear model, commonly called a meta-regression. The use of meta-regression models instead of a series of one-way ANOVA models helps reviewers avoid conducting too many statistical tests. However, computing the power of meta-regression models a priori is problematic. As seen in Hedges and Pigott (2004), we need to know the exact values of the moderators for each study in order to compute power for tests in meta-regression. These values cannot be guessed a priori, and thus, I will not present the details of these tests here.

References

Hedges, L.V., and T.D. Pigott. 2004. The power of statistical tests for moderators in meta-analysis. *Psychological Methods* 9(4): 426–445.

Higgins, J.P.T., and S.G. Thompson. 2002. Quantifying heterogeneity in a meta-analysis. *Statistics in Medicine* 21: 1539–1558.

Sirin, S.R. 2005. Socioeconomic status and academic achievement: A meta-analytic review of research. *Review of Educational Research* 75(3): 417–453. doi:10.3102/00346543075003417.

Chapter 7
Missing Data in Meta-analysis: Strategies and Approaches

Abstract This chapter provides an overview of missing data issues that can occur in a meta-analysis. Common approaches to missing data in meta-analysis are discussed. The chapter focuses on the problem of missing data in moderators of effect size. The examples demonstrate the use of maximum likelihood methods and multiple imputation, the only two methods that produce unbiased estimates under the assumption that data are missing at random. The methods discussed in this chapter are most useful in testing the sensitivity of results to missing data.

7.1 Background

All data analysts face the problem of missing data. Survey researchers often find respondents may refuse to answer a question, or may skip an item on a questionnaire. Experimental studies are also subject to drop-outs in both the treatment and control group. In meta-analysis, there are three major sources of missing data: missing studies from the review, missing effect sizes or outcomes for the analysis, and missing predictors for models of effect size variation. This chapter will provide strategies for testing the sensitivity of results to problems with missing data. As discussed throughout this chapter, current methods for missing data require strong assumptions about the reasons why data are missing, and about the distribution of the hypothetically complete data that cannot be verified empirically. Instead, re-analyzing the data under a number of differing assumptions provides the reviewer with evidence of the robustness of the results.

Over the past 20 years, statisticians have conducted an extensive amount of research into methods for dealing with missing data. Schafer and Graham (2002) point out that the main goal of statistical methods for missing data is not to recover or estimate the missing values but to make valid inferences about a population of interest. Thus, Schafer and Graham note that appropriate missing data techniques are embedded in the particular model or testing procedure used in the analysis. This chapter will take Schafer and Graham's perspective and provide missing data

T.D. Pigott, *Advances in Meta-Analysis*, Statistics for Social and Behavioral Sciences,   79
DOI 10.1007/978-1-4614-2278-5_7, © Springer Science+Business Media, LLC 2012

methods adapted from the statistical literature (Little and Rubin 1987; Schafer 1997) for use in meta-analysis. This chapter will focus on the sensitivity of results to missing data rather than on providing an alternative set of estimates that compensate for the missing data. For many missing data methods, the strategy involves recognizing the greater amount of uncertainty in the data caused by the missing information. Thus, many missing data methods result in a larger variance around the model estimates. This chapter will focus on methods that formally incorporate a larger amount of variance when missing data occurs.

The two most common strategies suggested for missing effect sizes in meta-analysis do not take into account the true level of uncertainty caused by missing data. These two strategies involve filling in either the observed mean using studies that report that missing variable, and filling in a zero for studies missing an effect size. Filling in the same value for missing observations in any data set will reduce the variance in the resulting estimate, making the estimates seem to contain more information than is truly available. Later in the chapter, we will discuss imputation strategies that incorporate a larger degree of uncertainty in the estimates reflecting the missing information in the data, and we will use these estimates to judge the sensitivity of results to assumptions about missing data.

7.2   Missing Studies in a Meta-analysis

One common form of missing data in a meta-analysis is missing studies. The most common cause of missing studies is publication bias. As many researchers have shown (Begg and Berlin 1988; Hemminki 1980; Rosenthal 1979; Smith 1980), there is a bias in the published literature toward statistically significant results. If a search strategy for a meta-analysis focuses only on published studies, then there is a tendency across many disciplines for the overall effect size to be biased toward statistically significant effects, thus over-estimating the true difference between the treatment and control group or the strength of the association between two measures. One strategy for addressing publication bias is the use of thorough search strategies that focus on published, unpublished and fugitive literatures. This section will provide an overview of strategies detecting and examining the potential for publication bias; more detailed information can be found in Rothstein et al. (2005).

7.2.1   Identification of Publication Bias

Even when a search strategy aims for a wide range of published and unpublished studies, the resulting sample of studies may still suffer from publication bias. One set of strategies associated with missing studies focuses on the identification of publication bias. The simplest and most widely known of these strategies is the funnel plot (Sterne et al. 2005). Funnel plots are scatterplots of the effect size on the

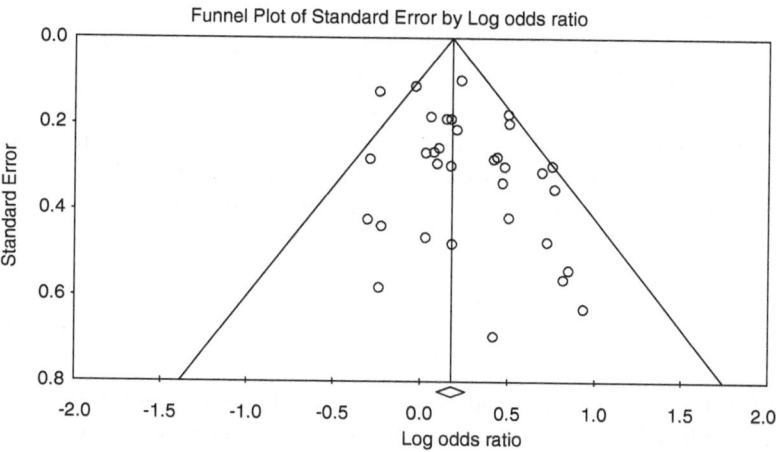

Fig. 7.1 Funnel plot for passive smoking data

x-axis and the sample size, variance, or study level weight of the effect size on the y-axis. (Recall that the study level weight is the inverse of the variance of the effect size). With no publication bias, the plot should resemble a funnel with the wider end of the funnel associated with studies of small sample sizes and large variances. The smaller end of the funnel should have effect sizes that have larger sample sizes and smaller variances, centered around the mean of the effect size distribution. If publication bias exists, then the plot will appear asymmetric. If small studies with small sample sizes are missing, then the funnel plot will appear to have a piece missing at its widest point. The quadrant with small effect sizes and small sample sizes would be those most likely to be censored in the published literature. If small effect sizes are, in general, more unlikely to appear in the literature, then the funnel will have fewer studies in the area of the graph corresponding to effect sizes close to zero, despite the sample size.

7.2.1.1   Example of Funnel Plot

Figure 7.1 is a funnel plot of data taken from Hackshaw et al. (1997) study of the relationship between passive smoking and lung cancer in women. The 33 studies in the meta-analysis compare the number of cases of lung cancer diagnosed in individuals whose spouses smoke with the number of cases of lung cancer in individuals whose spouses are non-smokers. The data used to construct this plot are given in the Data Appendix. The x-axis is the log-odds ratio, and the y-axis is the standard error of the log-odds ratio. There is a gap in the lower left-hand corner of the funnel plot, indicating that some studies with large standard errors and negative log-odds ratios could be missing. Thus, we see some evidence of publication bias here (Fig. 7.1).

While funnel plots are easily constructed, they can be difficult to interpret. Many conditions can lead to asymmetric plots even when the sample of studies is not affected by publication bias. A more formal test of publication bias was proposed by Egger et al. (1997) using regression techniques. Egger et al.'s method provides a test of whether a funnel plot shows evidence of asymmetry. The method involves standardizing the effect size into a standard normal deviate and regressing this transformed effect size on the precision of the effect size, defined as the inverse of the standard error of the effect size. The regression equation can be expressed as

$$\frac{T_i}{\sqrt{v_i}} = \hat{\beta}_0 + \hat{\beta}_1 \frac{1}{\sqrt{v_i}} \tag{7.1}$$

where T_i is the effect size for study i, and v_i is the standard error for the effect size in study i. When the funnel plot is symmetric, that is, when there is no evidence of publication bias, then $\hat{\beta}_0$ is close to zero. Symmetric funnel plots will produce an estimated regression equation that goes through the origin. Standardizing small effect sizes using the standard error should create a small standard normal deviate. In contrast, larger studies will produce larger standard normal deviates since their standard errors will be small. When publication bias is present, we may have large studies with normal deviates that are smaller than studies with small sample sizes – indicating that the small studies differ from large studies in their estimates of effect size.

Figure 7.2 provides the scatterplot of the standardized effect size by the inverse of the standard error of the effect size in the passive smoking data. The dotted line in the graph is the regression line given in (7.1). Table 7.1 provides the regression coefficients for (7.1) fit to the passive smoking data. The value for the intercept, β_0, is statistically different from zero, thus indicating that there is evidence of publication bias in the passive smoking data.

7.2.2   Assessing the Sensitivity of Results to Publication Bias

If a reviewer suspects publication bias in the meta-analytic data, there are two general classes of methods for exploring the sensitivity of results to publication bias. The first method, trim-and-fill (Duval and Tweedie 2000) is fairly easy to implement, but relies on strong assumptions about the nature of the missing studies. As Vevea and Woods (2005) point out, the trim-and-fill method assumes that the missing studies are one-to-one reflections of the largest effect sizes in the data set, in essence, that the missing studies have effect sizes values that are the negative of the largest effect sizes observed in the data set. In addition, the trim-and-fill method may lead to biased results if the effect sizes are in fact heterogeneous. Vevea and Woods present a second method that produces estimates for models of effect size under a series of possible censoring mechanisms, addressing the shortcomings they

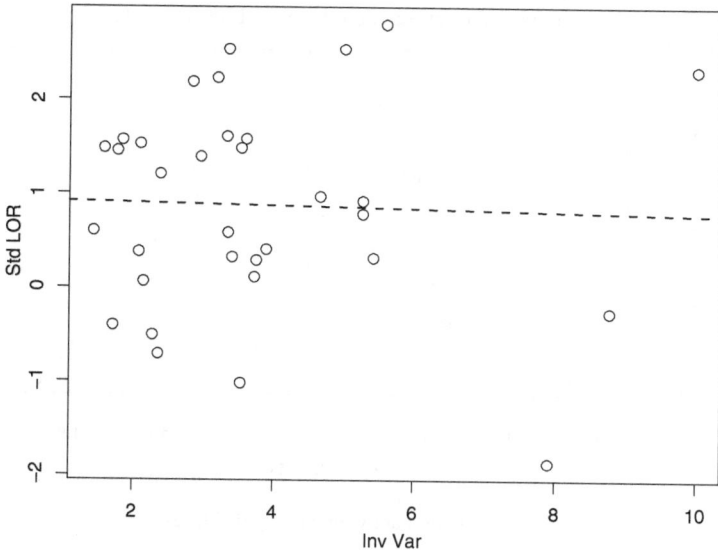

Fig. 7.2 Egger's test for publication bias for passive smoking data

Table 7.1 Egger's test for the passive smoking data set

Coefficient	Estimate	SE	t-Value	p
β_0	0.933	0.417	2.236	0.033
β_1	−0.0155	0.098	−0.158	0.875

find with the trim-and-fill method This second method provides more flexibility than trim-and-fill since it allows the examination of sensitivity for a range of models of effect size.

The trim-and-fill method (Duval and Tweedie 2000) is based on asymmetry in the funnel plot. This method builds on the funnel plot by "filling in" effect sizes that are missing from the funnel and then estimating the overall mean effect size including these hypothetical values. The assumption in this method is that there are a set of missing effect sizes that are "mirror images" of the largest effect sizes in the data set. In other words, the method inserts missing effect sizes that are of the opposite sign, and are mirror reflections of the largest observed effect sizes with similar sample sizes. The theoretical basis for the method is beyond the scope of this text, but the analysis itself is fairly simple. The idea is to first estimate how many effect sizes are "missing" in order to create a symmetric funnel plot. This computation may require a few iterations to obtain the estimate. Once the researcher computes the number of missing effect sizes, hypothetical values for these missing observations are used to re-estimate the mean effect size. This new mean effect size incorporates the possible effects of publication bias.

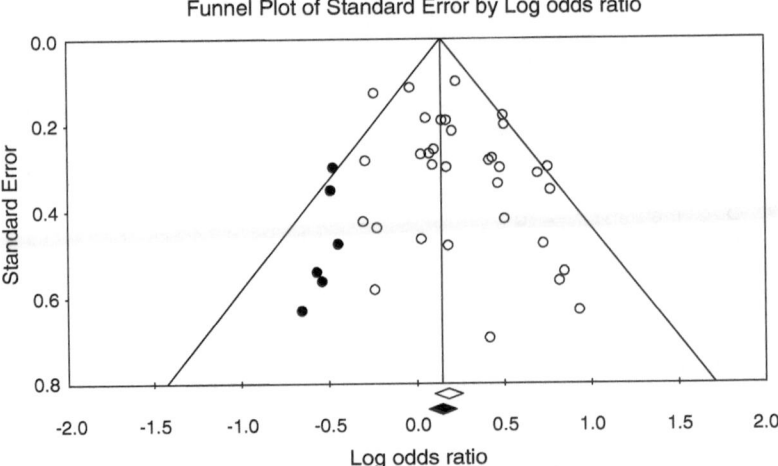

Fig. 7.3 Funnel plot for passive smoking data with Trim and Fill results

In practice, knowing how much bias is possible allows a test of the sensitivity of results. The reviewer should decide if the difference in these two values is of substantive importance, and should report these values to readers. While space does not permit a full example illustrating trim-and-fill, Duval (2005) provides a step-by-step outline for completing the method. Figure 7.3 is a funnel plot of the smoking data with the "missing" effect sizes represented by solid circles. The bottom of the plot shows the mean effect size computed with only the observed studies, and the mean effect size when the "missing" studies are included. As seen in Fig. 7.3, the mean effect size does not change significantly if we assume some publication bias.

A second strategy involves modeling publication bias using a censoring mechanism as described by Vevea and Woods (2005). The model illustrated by Vevea and Woods proposes a number of censoring mechanisms that could be operating in the literature. These censoring mechanisms are based on the type of censoring and the severity of the problem. In general, Vevea and Woods examine the impact of censoring where the smallest studies with the smallest effect sizes are missing (one-tailed censoring) and where the studies with non-significant effect sizes (two-tailed censoring) are missing. Vevea and Woods point out that the trim-and-fill method assumes that the missing studies are one-to-one reflections of statistically significant effect sizes, and that the method can only examine the sensitivity of estimates of the mean effect size. Vevea and Woods' method can examine the sensitivity of fixed, random and mixed effects models of effect size to publication bias. These methods do require the use of more flexible computing environments, and reviewers may find them more difficult. Readers interested in methods for publication bias will find more details in Rothstein et al. (2005).

7.3   Missing Effect Sizes in a Meta-analysis

When effect sizes are missing from a study, there are few missing data strategies to analyze the data. The most common method used by reviewers is to drop these studies from the analysis. The problem is similar to missing data in primary studies. If an individual patient does not have a measure for the target outcome, then that patient cannot provide any information about the efficacy of the treatment.

One reason for missing effect sizes is that reviewers either cannot compute an effect size from the information given in a study or do not know how to compute an effect size from the results of the study. For example, studies may fail to report summary statistics needed to compute an effect size such as means and standard deviations for standardized mean differences, or frequencies for the odds ratio. Other studies may report only the summary test statistic such as a t-test or an F-test, or only the p-value for the test. These difficulties often occur with older studies since professional organizations such as the American Psychological Association and the American Medical Association now have standards for reporting that assist reviewers in extracting information from a meta-analysis.

Another reason for missing effect sizes in more recently published studies arises when a reviewer does not know how to translate complex statistical results derived from techniques such as factorial ANOVA, regression models, or hierarchical linear models into an effect size for the review. A related problem occurs when the studies in the review report a wide variety of statistics. For example, one study in the review may report a t-test from a quasi-experimental study, while another study may report a correlation coefficient from an observational study. The question is whether these differing measures of effect size can and should be combined in the meta-analysis, or whether the reviewer should consider certain types of effect sizes as missing. Lipsey and Wilson (2001) and Shadish et al. (1999) provide a number of tools for computing effect sizes from commonly provided information in a study. Wilson (2010) provides a free, web-based, effect size calculator that obtains effect sizes from a large array of reported statistics. In practice, the reviewer should try multiple methods for computing an effect size from the data given in the study, and to contact primary authors for available information. If effect sizes are still not available, then the reviewer should explore the sensitivity of results to publication bias. One suggestion often used by reviewers is to impute a value of 0 for all missing effect sizes. While this method seemingly provides a conservative estimate of the mean effect size, the standard errors for the mean effect size will be underestimated. As will be discussed in the next sections, imputing a single value for any missing observation will not reflect accurately the uncertainty in the data.

A second reason for missing outcomes in a study is selective reporting. A number of researchers in medicine have documented primary studies where researchers have not reported on an outcome that was gathered. In many cases, these outcomes are also ones that are not statistically significant, or may be outcomes reporting on an adverse outcome. In the Cochrane Collaboration, reviewers are required to report on whether selective reporting of outcomes has occurred in a study in the Risk of

Bias table. Reviewers who suspect outcome reporting bias have few strategies for dealing with this problem aside from contacting the primary authors for the missing information. Researchers (Chan et al. 2004; Williamson et al. 2005) have also developed methods for assessing the potential bias in results when some outcomes are selectively missing. In the long term, reporting standards and registries of primary studies may be the most effective strategies for ensuring complete reporting of all outcomes gathered in a study.

7.4   Missing Moderators in Effect Size Models

Another form of missing data in a meta-analysis is missing moderators or predictors in effect size models. This form of missing data occurs when the reviewer wants to code particular aspects of the primary study, but this information is not provided in the study report. A reviewer examining a body of research on a topic has, in one sense, fewer constraints than a primary researcher. When a primary researcher plans a study, they must make decisions about the number and type of measures used, the optimal design and methods, and in the final report, about what information is most relevant. In contrast, a reviewer can examine questions about how the different decisions made by the primary researcher about study design, measures, and reporting relates to variation in study results.

Disciplinary practices in a given area of research may constrain how primary authors collect, measure and report information in a study. Orwin and Cordray (1985) use the term macrolevel reporting to refer to practices in a given research area that influence how constructs are defined and reported. In a recent meta-analysis, Sirin (2005) found multiple measures of socioeconomic status in studies examining the relationship between income and academic achievement. These measures included ratings of parents' occupational status, parental income, parental education, free lunch eligibility, as well as composites of these measures. Researchers in some fields may be more inclined to use ratings of parents' occupation status whereas other educational researchers may only have a measure of free lunch eligibility available. Thus, parents' occupation status may be missing in some studies since the researchers in one field may rely on free lunch eligibility, for example, as the primary measure of socioeconomic status. Differences in reporting among primary studies could be related to disciplinary practices in the primary author's field. Primary authors may also be constrained by a particular journal's publication practices, and thus do not report on information a reviewer may consider important.

Researchers also differ in their writing styles and thoroughness of reporting. Orwin and Cordary (1985) use the term microlevel reporting quality to refer to individual differences among researchers in their reporting practices. In some cases, a moderator that is not reported among all studies in a review could be missing in a random way due to the individual differences among researchers.

Another reason study descriptors may be missing from primary study reports relates to bias for reporting only statistically significant results. For example, a primary researcher may be interested in the relationship between the percentage of low-income children in a classroom and classroom achievement. If the primary researcher finds that the percentage of low-income children in a classroom is not related to classroom achievement, then they may report only that the statistic did not reach statistical significance. Williamson et al. (2005) provide a discussion of this problem raising the possibility that the likelihood of a study reporting a given descriptor variable is related to the statistical significance of the relationship between the variable and the outcome. In this case, we have selective predictor reporting that operates in a similar manner to selective outcome reporting as described above.

Researchers may also avoid reporting study descriptors when those values are not generally acceptable. For example, if a researcher obtains a racially homogeneous sample, the researcher may hesitate to report fully the actual distribution of ethnicity in the sample. This type of selective reporting may also occur when reporting in a study is influenced by a desire to "clean up" the results. For example, attrition information may not be reported in a published study in order to present a more positive picture. In both of these cases, the values of the study descriptor influence whether the researcher reports those values.

Missing data in a meta-analysis occur in the form of missing studies from the review, missing effect sizes and missing predictors. The cases of missing studies and missing effect sizes correspond to missing an outcome in a primary study. The only approach to missing outcomes is checking the sensitivity of results to publication bias or censoring. The major class of missing data methods in the statistical literature applies most directly to missing data on predictors in models of effect size. The remainder of this chapter discusses the assumptions, techniques and interpretations of missing data methods applied to meta-analysis with an emphasis on handling missing predictors in effect size models.

7.5   Theoretical Basis for Missing Data Methods

The general approach used in current missing data methods involves using the data at hand to draw valid conclusions, and not to recover all the missing information to create a complete data set. This approach is especially applicable to meta-analysis since missing data frequently occur because a variable was not measured, and is not recoverable. Thus, the approaches taken in this chapter do not attempt to replace individual missing observations, but instead either estimate the value for summary statistics in the presence of missing data or sample several possible values for the missing observations from a hypothetical distribution based on the data we do observe.

The methods used in this chapter make strong assumptions about the distribution of the data, and about the mechanism that causes the missing observations.

Generally, the methods here require the assumption that the joint distribution of the effect size and moderator variables is multivariate normal. A second assumption is that the reasons for the missing data do not depend on the values of the missing observations. For example, if our data are missing measures of income for studies with a large proportion of affluent participants, then the methods we discuss here could lead to biased estimates. One major difficulty in applying missing data methods is that assumptions about the nature of the missing data mechanism cannot be tested empirically. These assumptions can only be subjected to the "is it possible" test, i.e., is it possible that the reasons for missing observations on a particular variable do not depend directly on the values of that variable? Missing observations on income usually fail the test, since it is a well-known result in survey sampling that respondents with higher incomes tend not to report their earnings. The following section examines these assumptions in the context of meta-analysis.

7.5.1　Multivariate Normality in Meta-analysis

The missing data methods used in this chapter rely on the assumption that the joint distribution of the data is multivariate normal. Thus, meta-analysts must assume that the joint distribution of the effect sizes and the variables coded from the studies in the review follow a normal distribution. One problematic issue in meta-analysis concerns the common incidence of categorical moderators in effect size models. Codes for characteristics of studies often take on values that indicate whether a primary author used a particular method (e.g., random assignment or not) or a certain assessment for the outcome (e.g., standardized protocol or test, researcher developed rating scale, etc.). Schafer (1997) indicates that in the case of categorical predictors, the normal model will still prove useful if the categorical variables are completely observed, and the variables with missing observations can be assumed multivariate normal conditional on the variables with complete data. The example later in the chapter examines a meta-analysis on gender differences in transformational leadership. Say we have missing data from some studies on the percent of men in the sample of participants who are surveyed in each study. We can still fulfill the multivariate normality condition if we can assume that the variable with missing observations, the percent of males in the sample, is normally distributed conditional on a fully observed categorical variable such as whether or not the first author is a woman. If the histogram of the percent of male subjects in the sample is normally distributed for the set of studies with male first authors and for the set of studies with female first authors, then we have not violated the normality assumption. Some ordered categorical predictors can also be transformed to allow the normal assumption to apply. If key moderators of interest are non-ordered categorical variables, and these variables are missing observations, then missing data methods based on the multinomial model may apply. There is currently no research on how to handle missing categorical predictors in meta-analysis.

7.5.2   Missing Data Mechanisms or Reasons for Missing Data

In addition to assuming that the joint and/or conditional distribution of the data is multivariate normal, the methods discussed in this chapter also require the assumption that the missing data mechanism is ignorable. There are two conditions that meet the conditions of ignorability, missing completely at random (MCAR) data, and missing at random (MAR) data (Rubin 1976). Missing data are missing completely at random when the cases with completely observed data are a random sample of the original data. When there are small amounts of missing data on an important variable, often analysts assume that the completely observed cases are as representative of the target population as the original sample. Thus, when data are missing completely at random, the data analyst does not need to know the exact reasons or mechanisms that caused the missing data; analyzing only the cases with complete observations will yield results that provide unbiased estimates of population parameters.

As applied to meta-analysis, individual differences among primary authors in their reporting practices may result in missing predictors that could be considered missing completely at random. The difficulty lies in gathering evidence that the missing predictors are missing completely at random. One strategy suggested is using logistic regression models to examine the relationships between whether a given predictor is observed or not and values of other variables in the data set. The difficulty arises when these models do not adequately explain the probability of missing a predictor. The relationships between observed variables and missing variables could be more complex than represented in the logistic regression model, or the probability of observing a value could be dependent on other unknown information. Schafer (1997) suggests that a more practical solution is to use as much information in the data set to estimate models in the presence of missing data, a point that will be elaborated later in the chapter.

A second condition that meets the conditions of ignorability is data missing at random. Unlike MCAR data, the cases with completely observed data are not a random sample of the original data. When data are MAR, the probability of missing an observation depends on the values of completely observed variables. Data are MAR if the conditional distribution of the missing values depends only on completely observed variables and not on variables with missing observations. This assumption is less stringent than MCAR, and is plausible in many missing data situations in meta-analysis. For example, some studies may report the income level of subjects as a function of the percent of students who qualify for free lunch, while others report income level as the average income reported by parents. The differences between these studies could be due to the discipline of the primary author – studies in education tend to use the percent of students with low income in a school while larger-scale studies may have the resources to conduct a survey of parents to obtain a more direct measure of income. A missing value for a particular measure of income in a particular study is not necessarily related to the value of income itself but to the choices of the primary author and constraints on the published version of the study. Thus, if we can posit a plausible mechanism for the missing observations, say the disciplinary background of the primary author, and this variable is completely observed in the data, then we can consider income in this instance MAR.

One set of methods that cannot be addressed fully in this chapter are methods for nonignorable missing data. Nonignorable missing data occur when the reason for a missing observation is the value of that observation. For example, self-reports of income are more frequently missing for those respondents with high income, a classic example of nonignorable missing data. In the case of nonignorable missing data, the analysis must incorporate a model for the missing data mechanism instead of ignoring it as in the case of MCAR and MAR data. A special case of nonignorable missing data is publication bias. As discussed earlier in the chapter, these methods usually require specialized computing environments, and provide information about the sensitivity of results to assumptions about the missing data.

7.6 Commonly Used Methods for Missing Data in Meta-analysis

Prior to Little and Rubin's (1987) work, most researchers employed one of three strategies to handle missing data: using only cases with all variables completely observed (listwise deletion), using available cases that have particular pairs of variables observed (pairwise deletion), or replacing missing values on a given variable with a single value such as the mean for the complete cases (single value imputation). The performance of these methods depends on the validity of the assumptions about why the data are missing. In general, these methods will produce unbiased estimates only when data are missing completely at random. The main problem with the use of these methods is that the standard errors do not accurately reflect the fact that variables have missing observations.

7.6.1 Complete-Case Analysis

In complete-case analysis, the researcher uses only those cases with all variables fully observed. This procedure, listwise deletion, is usually the default procedure for many statistical computer packages. When the missing data are missing completely at random, then complete-case analysis will produce unbiased results since the complete cases can be considered a random sample from the originally identified set of cases. Thus, if a synthesist can make the assumption that values are missing completely at random, using only complete cases will produce unbiased results.

Table 7.2 Complete case analysis using the gender and leadership data

Variable	Coefficient	SE	Z	p
Intercept	58.033	22.650	2.562	0.005
Publication year	−0.028	0.011	−2.446	0.007
Average age of sample	−0.040	0.005	−7.946	0.000
Percent of male leaders	0.001	0.002	4.472	0.000
First author female	−0.372	0.087	−4.296	0.000
Size of organization	−0.308	0.103	−2.989	0.001
Random selection used	0.110	0.035	3.129	0.000

In meta-analysis as in other analyses, using only complete cases can seriously limit the number of observations available for the analysis. Losing cases decreases the power of the analysis, and also ignores the information contained in the incomplete cases (Kim and Curry 1977; Little and Rubin 1987).

When data are missing because of a nonignorable response mechanism or are MAR, complete case analysis yields biased estimates since the complete cases cannot be considered representative of the original sample. With nonignorable missing data, the complete cases observe only part of the distribution of a particular variable. With MAR data, the complete cases are also not a random sample of the original sample. But with MAR data, the incomplete cases still provide information since the variables that are completely observed across the data set are related to the probability of missing a particular variable.

7.6.1.1 Example: Complete-Case Analysis

The data for the examples that follow are adapted from a meta-analysis by Eagly et al. (2003) that examines gender differences in transformational, transactional, and laissez-faire leadership styles. Six moderators are used in this example: Year the study was published (Year), if the first author is female, the average age of the participants, the size of the organization where the study was conducted, whether random methods were used to select the participants from the organization, and the percent of males in the leadership roles in the organization. Table 7.2 provides the complete case results for a meta-regression of the transformational leadership data. Only 22 or 50% of the cases include all five predictor variables. Positive effect sizes indicate that males were found to score higher on transformational leadership scales, while negative effect sizes favor females. For this sample of studies, gender differences in favor of females are associated with more recently published studies, with studies that have older samples of participants, when the study's first author is female, and in studies conducted in larger organizations. Gender differences in favor of males are found in studies with a higher percentage of men in leadership roles in the sample, and when random methods were used to select the sample from the target population. The complete case results can be obtained from any weighted regression program usually by default. Most statistical computing packages automatically delete any cases missing at least one variable in the model.

Table 7.3 Missing data patterns in leadership data

Pattern	Effect size	% male leaders	Average age	N
0	O	O	O	22 (50%)
1	O	O	–	15 (34%)
2	O	–	O	3 (7%)
3	O	–	–	4 (9%)
N	44 (100%)	37 (84%)	25 (57%)	44 (100%)

O indicates observed in data

7.6.2   *Available Case Analysis or Pairwise Deletion*

An available-case analysis or pairwise deletion estimates parameters using as much data as possible. In other words, if three variables in a data set are missing 0%, 10% and 40% of their values, respectively, then the correlation between the first and third variable would be estimated using the 60% of cases that observe both variables, and the correlation between the first and second variable would use the 90% of cases that observe these two variables. A different set of cases would provide the estimate of the correlation of the second and third variables since there is the potential of having between 40% and 50% of these cases missing both variables. For example, Table 7.3 shows the pairs of cases that would be used to estimate parameters using available case analysis in the leadership data. The letter O indicates that the variable was observed in that missing data pattern. Assuming the effect sizes are completely observed, the estimate of the correlation between the effect size and the percentage of men in leadership roles would be based on 37 studies or 84% of the studies. The correlation between effect size and average age of the sample would use 25 studies or 57% of the sample. The estimated correlation of percentage of men in leadership roles and average age of the sample would use only 22 studies or 50% of the sample.

This simple example illustrates the drawback of using available case analysis – each correlation in the variance-covariance matrix estimated using available cases could be based on different subsets of the original data set. If data are MCAR, then each of these subsets are representative of the original data, and available case analysis provides estimates that are unbiased. If data are MAR, however, then each of these subsets is not representative of the original data and will produce biased estimates.

Much of the early research on methods for missing data focuses on the performance of available case analysis versus complete case analysis (Glasser 1964; Haitovsky 1968; Kim and Curry 1977). Fahrbach (2001) examines the research on available case analysis and concludes that available case methods provide more efficient estimators than complete case analysis when correlations between two independent variables are moderate (around 0.6). This view, however, is not shared by all who have examined this literature (Allison 2002).

One statistical problem that could arise from the use of available cases under any form of missing data is a non-positive definite variance-covariance matrix, i.e., a

Table 7.4 Available case analysis

Variable	Coefficient	SE	Z	p
Intercept	−46.283	19.71	−2.348	0.009
Publication year	0.024	0.010	2.383	0.009
Average age of sample	−0.047	0.006	−7.625	0.000
Percent of male leaders	0.013	0.002	6.166	0.000
First author female	−0.260	0.058	−4.455	0.000
Size of organization	0.185	0.062	3.008	0.001
Random selection used	0.037	0.041	0.902	0.184

variance-covariance matrix that cannot be inverted to obtain the estimates of slopes for a regression model. One reason for this problem is that different subsets of studies are used to compute the elements of the variance-covariance matrix. Further, Allison (2002) points out that a more difficult problem in the application of available case analysis concerns the computation of standard errors of available case estimates. At issue is the correct sample size when computing standard errors since each parameter could be estimated with a different subset of data. Some of the standard errors could be based on the whole data set, while others may be based on the subset of studies that observe a particular variable or pair of variables. Though many statistical computing packages implement available case analysis, how standard errors are computed differs widely.

7.6.2.1   Example: Available Case Analysis

Table 7.4 provides the results from SPSS estimating a meta-regression for the leadership studies using pairwise deletion. In this analysis, gender differences favoring females are associated with an older sample and with studies whose first author is female. Gender differences favoring males are associated with a higher percentage of men in leadership roles, larger organizations in the sample, and with more recent studies. The last two findings, related to larger organizations and more recent findings, contradicts the findings from the complete case analysis.

While available case analysis is easy to understand and implement, there is little consensus in the literature about the conditions where available case analysis outperforms complete case analysis when data are MCAR. As described above, the performance of available case analysis may relate to the size of the correlations between variables in the data, but there is no consensus about the optimal size of these correlations needed to produce unbiased estimates.

7.6.3   Single Value Imputation with the Complete Case Mean

When values are missing in a meta-analysis (or in any statistical analysis), many researchers have replaced the missing value with a "reasonable" value such as the

Table 7.5 Comparison of complete-case and mean imputation values

Variable	N	Mean	SD
Average age of sample	22	44.88	6.629
Average age, mean imputed	*44*	*44.88*	*4.952*
Percent of male leaders	22	64.97	17.557
Percent of male leaders, mean imputed	*44*	*64.97*	*16.060*

Table 7.6 Linear model of effect size for leadership data using mean imputation

Variable	Coefficient	SE	Z	p
Intercept	21.165	11.468	1.846	0.032
Publication year	−0.01	0.006	−1.783	0.037
Average age of sample	−0.024	0.004	−6.537	0.000
Percent of male leaders	0.006	0.001	4.753	0.000
First author female	−0.076	0.035	−2.204	0.014
Size of organization	−0.064	0.034	−1.884	0.030
Random selection used	−0.013	0.028	−0.469	0.319

mean for the cases that observed the variable. Little and Rubin (1987) refer to this strategy as single-value imputation. Researchers commonly use two different strategies to fill in missing values. One method fills in the complete case mean, and the other uses regression with the complete cases to estimate predicted values for missing observations given the observed values in a particular case.

Replacing the missing values in a variable with the complete case mean of the variable is also referred to as unconditional mean imputation. When we substitute a single value for all the missing values, the estimate of the variance of that variable is decreased. The estimated variance thus does not reflect the true uncertainty in the variable – instead the smaller variance wrongly indicates more certainty about the value. These biases get compounded when using the biased variances to estimate models of effect size. Imputation of the complete case mean never leads to unbiased variances of variables with missing data.

7.6.3.1   Example: Mean Imputation

In Table 7.5, the missing values of average age and percent of men in leadership roles were imputed with the complete case mean. These values are given under the complete case means and standard deviations. While the means for the variables remains the same, the standard deviations are smaller for the variables when missing values are replaced by the complete case mean. The problem is compounded in the regression analysis in Table 7.6. These results would lead us to different conclusions from those based on either the complete-case or pairwise deletion analyses.

Using mean imputation in this example would lead us to conclude that men score higher on transformational leadership scales only in studies that have a larger percentage of male leaders in the sample. The use of random selection is not related to variability in the effect size across studies. All the other predictors favor women's scores on transformational leadership.

7.6.4   Single Value Imputation Using Regression Techniques

A single-value imputation method that provides less biased results with missing data was first suggested by Buck (1960). Instead of replacing each missing value with the complete case mean, each missing value is replaced with the predicted value from a regression model using the variables observed in that particular case as predictors and the missing variable as the outcome. This method is also referred to as conditional mean imputation or as regression imputation. For each pattern of missing data, the cases with complete data on the variables in that pattern are used to estimate regressions using the observed variables to predict the missing values. The end result is that each missing value is replaced by a predicted value from a regression using the values of the observed variables in that case. When data are MCAR, then each of the subsets used to estimate prediction equations are representative of the original sample of studies. This method results in more variation than in unconditional mean imputation since the missing values are replaced with values that depend on the regression equation. However, the standard errors using Buck's method are still too small. This underestimation occurs since Buck's method replaces the missing values with predicted values that lie directly on the regression line used to impute the values. In other words, Buck's method results in imputing values that are predicted exactly by the regression equation without error.

Little and Rubin (1987) present the form of the bias for Buck's method and suggest corrections to the estimated variances to account for the bias. If we have two variables, Y_1 and Y_2, and Y_2 has missing observations, then the form of the bias using Buck's method to fill in values for Y_2 is given by

$$(n - n^{(2)})(n - 1)^{-1}\sigma_{22.1},$$

where n is the sample size, $n^{(2)}$ is the number of cases that observe Y_2, and $\sigma_{22.1}$ is the residual variance from the regression of Y_2 on Y_1. Little and Rubin also provide the more general form of the bias with more than two variables. Table 7.7 compares the complete case means and standard deviations of average age of the sample and percent of male leaders with those obtained using regression imputation (Buck's method) and Little and Rubin (1987) correction.

The standard deviations for the corrected regression imputation results are larger than for both complete cases and for the uncorrected regression imputation. The correction reflects the increased uncertainty in the estimates due to missing data.

Note that correcting the bias in Buck's method involves adjusting the variance of the variable with missing observations. Using Little and Rubin (1987) correction

Table 7.7 Comparison of methods for imputing missing data

Variable	Complete cases N = 22		Regression imputation (uncorrected) N = 44		Regression imputation (corrected) N = 44	
	Mean	SD	Mean	SD	Mean	SD
Average age of sample	45.23	6.70	44.52	5.52	44.52	6.91
Percent of male leaders	64.23	17.44	65.44	16.22	65.44	17.52

results in a corrected covariance matrix, and not individual estimates for each missing observation. Thus, estimating the linear model of effect size in our example will require estimating the coefficients using only the variance-covariance matrix.

To date, there has not been extensive research on the performance of Buck's method to other more complex methods for missing data in meta-analysis. As seen above, one advantage of using Buck's method with the corrections suggested by Little and Rubin is that the standard errors of estimates reflect the uncertainty in the data and lead to more conservative and less biased estimates than complete case and available case methods. While using Buck's method is fairly simple, the adjustments of the variances and covariances of variables with missing observations adds another step to the analysis. In addition, it is not clear how to utilize the corrected variances and covariances when estimating weighted regression models of effect sizes in meta-analysis. While it is possible to estimate a linear model using a covariance matrix in the major statistical packages, it is not clear how to incorporate the weights into the corrected covariance matrix.

When missing predictors are MCAR, then complete case analysis, available case analysis and conditional mean imputation have the potential for producing unbiased results. The cost of using these methods lies in the estimation of standard errors. For complete case analysis, the standard errors will be larger than those from the hypothetically complete data. In available case and conditional mean imputation, the standard errors will be too small, though those from conditional mean imputation can be adjusted. When data are MAR or are missing due to a nonignorable response mechanism, none of the simpler methods produce unbiased results.

7.6.4.1 Example: Regression Imputation

Table 7.8 provides the estimates of the linear model of effect size when using regression imputation. The results in the table were produced using SPSS Missing Values Analysis, saving a data set where missing values are imputed using regression.

These results differ from the mean imputation results in that both when the first author is female and whether random selection was used are not related to effect size magnitude. As in the mean imputation results, only percent of male leaders is related to high scores on transformational leadership for men. These results are again not consistent with the available case or complete case results.

Table 7.8 Linear model of effect size for leadership data using regression imputation

Variable	Coefficient	SE	Z	p
Intercept	38.436	11.181	3.438	0.001
Publication year	−0.019	0.005	−3.515	<0.001
Average age of sample	−0.024	0.003	−7.105	<0.001
Percent of male leaders	0.008	0.001	5.921	<0.001
First author female	−0.044	0.035	−1.252	0.105
Size of organization	−0.079	0.034	−2.339	0.010
Random selection used	−0.018	0.028	−0.65	0.258

7.7   Model-Based Methods for Missing Data in Meta-analysis

The simple methods for missing data discussed above do not provide unbiased estimates under all circumstances. The general problem with these ad hoc methods is that they do not take into account the distribution of the hypothetically complete data. For example, filling in a zero for a missing effect size may be a reasonable assumption, but it is not based on a plausible distribution for the effect sizes in a review. The missing data methods outlined in this section begin with a model for the observed data. Maximum-likelihood methods using the EM algorithm are based on the observed data likelihood while multiple imputation techniques are based on the observed data posterior distribution. Given the assumptions of ignorable missing data and multivariate normal data, the observed data likelihood and the observed data posterior distribution will provide the information needed to estimate important data parameters. The next sections outline both methods, providing an example of its application.

7.7.1   Maximum-Likelihood Methods for Missing Data Using the EM Algorithm

In statistical inference, we are interested in obtaining an estimate of a parameter from our data that has optimal properties such as minimum variance and bias. The method most often used to obtain parameter estimates is maximum likelihood. Maximum likelihood methods are based on a joint density distribution of the data. For example, if we assume that our data consist of a sample of n observations from the normal distribution, then we can write down our joint density as a product of a series of n normal densities. The maximum value for this density function is a parameter estimate that has optimal properties. For example, the arithmetic average of a set of observations from a normal distribution is the maximum likelihood estimate of the mean of the population.

When the data include missing observations, our data likelihood becomes complicated. Since our goal is to make inferences about the population, the relevant likelihood for our problem is the hypothetically complete data likelihood. This complete data likelihood includes the density of the observed data given the unknown parameters of the population distribution and the density of the missing data given the observed data and the unknown population parameters. Since we do not know the density of the missing data, it would seem impossible to compute the maximum likelihood estimates of the complete data. However, Dempster et al. (1977) developed an algorithm called the Expectation-Maximization algorithm, or EM algorithm. As its name denotes, obtaining maximum likelihood estimates requires an iterative process. In the first step, the Expectation or E-step, the algorithm uses an estimate of the data parameters (such as the mean and covariance matrix) to estimate plausible values for the missing observations. These missing observations are then "filled in" for the Maximization or M-step where the algorithm re-estimates the population parameters. Thus, in the E-step, we assume that we have estimates of the population parameters, and we use these to obtain values for the missing observations. In the M-step, we assume that the missing observations are "real" and use them to re-compute the population parameters. The iterations continue until the estimates of the population parameters do not change, i.e., when the algorithm converges.

When we can assume the data is multivariate normal, the distribution for the missing values given the observed values is also multivariate normal. It is important to note that the algorithm provides the maximum likelihood estimates of the sufficient statistics, the means and covariance matrix, and not the maximum likelihood of any particular missing observation. Thus, we have maximum likelihood estimates of the means, variances and covariances of our data that we can use to obtain estimates of other parameters such as regression coefficients for a linear model. We must also assume that the response mechanism is ignorable, i.e., that our data are either MCAR or MAR so that the distribution of our data does not need to include a specification of the response mechanism.

7.7.1.1 Example Using the EM Algorithm

There are a number of options available in commercially-available statistical packages and in freeware for use in computing the EM estimates. Using Schafer's NORM (1999) program, we obtain the maximum likelihood estimates of the means and standard deviations for our two variables with missing data, average of the sample and percent of male leaders as seen in Table 7.9. These estimates are compared with the estimates from the complete case and corrected regression imputation analyses. The standard deviation from the EM algorithm falls between the complete case and regression imputation estimates. The standard deviations are of the same magnitude as the complete case estimates, generally reflecting the same amount of "information" as in complete cases.

Table 7.9 Comparison of estimates from the EM algorithm, complete-case analysis and regression imputation

Variable	Complete cases N = 22		Regression imputation (corrected) N = 44		EM algorithm N = 44	
	Mean	SD	Mean	SD	Mean	SD
Average age of sample	45.23	6.70	44.52	6.91	44.44	6.67
Percent of male leaders	64.23	17.44	65.44	17.52	65.65	17.21

The difficulty with using the EM algorithm in the context of meta-analysis is similar to that of the corrected regression imputation analysis – it is not clear how to estimate the weighted regression coefficients using the sufficient statistics matrix (the matrix of means, variances and covariances). Thus, the EM algorithm has limited applicability to meta-analysis since we do not yet have a method for computing the weighted regression coefficients from the means, variances and covariances of the variables in the data.

7.7.2   Multiple Imputation for Multivariate Normal Data

Multiple imputation has become the method of choice in many contexts of missing data. The main advantage of multiple imputation is that the analyst uses the same statistical procedures in the analysis phase that were planned for completely observed data. In other words, in the analysis phase of multiple imputation, the researcher does not need to adjust standard errors as in Buck's method, and does not need to estimate a regression from the covariance matrix as in maximum likelihood with the EM algorithm. Multiple imputation, as its name implies, is a technique that generates multiple possible values for each missing observation in the data. Each of these values is used in turn to create a complete data set. The analyst uses standard statistical procedures to analyze each of these multiply imputed data sets, and then combines the results of these analyses for statistical inference.

Thus, multiple imputation consists of three phases. The first phase consists of generating the possible values for each missing observation. The second phase then analyzes each completed data set using standard statistical procedures. The third phase involves combining the estimates from the analyses of the second phase to obtain results to use for statistical inference. Each of these phases is discussed conceptually below. Readers interested in more details should consult the following works (Enders 2010; Schafer 1997).

7.7.2.1   Generating Multiple Imputations

Multiple imputation, like maximum likelihood methods for missing data, relies on a model for the distribution of missing data given the observed data under the condition of MAR data. As in maximum likelihood, this distribution is complex. The previous section discussed the use of the EM algorithm to estimate sufficient statistics from this distribution assuming that the hypothetically complete data is multivariate normal. Multiple imputation uses Bayesian methods to obtain random draws from the posterior predictive distribution of the missing observations given the observed observations. These random draws are completed in an iterative process much like the EM algorithm. Given the means and covariance matrix of our hypothetically complete multivariate normal data, we can then obtain the form of the distribution of the missing observations given the observed data, and draw a random observation from that distribution. That observation would be one plausible value for a missing value for a given case. Once we have drawn plausible values for all our missing observations, we obtain a new estimate of our means and covariance matrix, and repeat the process. Note that again we are assuming that our response mechanism is ignorable so that the posterior distribution also does not include a specification of the response mechanism.

In order to generate these random draws, however, we need to use simulation techniques such as Markov Chain Monte Carlo. These methods allow the use of simulation to obtain random draws from a complex distribution. While this phase is the most complex statistically, there are many commercial software packages and freeware available to generate these imputations, especially in the case where we can assume the complete data is multivariate normal. The Appendix provides details about these computer packages.

7.7.2.2 Analyzing the Completed Data Sets

Multiple imputation was first developed in large-scale survey research to assist researchers who wanted to use public data sets. The idea was to provide researchers with a way to handle missing data that did not require specialized computer programming skills or statistical expertise when using these publicly-available data sets. In this second step, the researcher will obtain a series of completed data sets, with each missing observation filled in using the methods in the prior section. Once the imputations are generated, the analyst uses whatever methods were originally planned for the data. These analyses are repeated for each completed data set. As Schafer (1997) argues, for most applications of multiple imputation, five imputations is sufficient to obtain estimates for statistical inference. In this phase, the analyst takes each completed data set and obtains estimates for the originally planned model. Table 7.10 provides the estimates for the linear model of effect size for each of five imputations generated as discussed in the Appendix.

7.7.2.3 Combining the Estimates

Rubin (1987) provides the formulas for combining the multiply-imputed estimates to obtain overall estimates and their standard errors. Let us denote the mean of our target estimate for the ith parameter across all j imputations as

Table 7.10 Imputations for the leadership data

Variable	Imputation 1	Imputation 2	Imputation 3	Imputation 4	Imputation 5
Intercept	17.219	3.276	31.380	0.429	54.661
Publication year	−0.008	−0.001	−0.016	0.000	−0.027
Average age of sample	−0.018	−0.026	−0.010	−0.021	−0.031
Percent of male leaders	0.005	0.004	0.003	0.007	0.009
First author female	−0.174	−0.186	−0.004	−0.147	−0.181
Size of organization	−0.016	0.015	−0.046	−0.021	0.063
Random selection used	0.047	0.068	−0.043	−0.018	0.112

Table 7.11 Multiply-imputed regression coefficients

Variable	Coefficient	SE	Z	p
Intercept	21.393	27.726	0.77	0.29
Publication year	−0.010	0.014	−0.75	0.30
Average age of sample	−0.021	0.009	−2.34	0.13
Percent of male leaders	0.006	0.003	2.06	0.14
First author female	−0.138	0.094	−1.47	0.19
Size of organization	−0.001	0.059	−0.02	0.49
Random selection used	0.033	0.076	0.43	0.37

$$\bar{q}_i = \frac{1}{m} \sum_{j=1}^{m} q_{ij}$$

where q_{ij} is the estimate of the ith parameter from the jth completed-data sets. To obtain the standard errors of the q_{ij}, we need two estimates of variance. Denote the variance of the estimate, q_{ij}, from the jth completed sets as $se^2(q_i)$. The variance across the j data sets of the estimate q_i is given by

$$s_{q_i}^2 = \frac{1}{m-1} \sum_{j=1}^{m} (q_{ij} - \bar{q}_i)^2.$$

The standard error of the points estimates of the q_{ij} is then given by

$$SE(q_i) = \sqrt{\frac{1}{m} \sum_{j=1}^{m} se^2(q_i) + s_{q_i}^2 \left[1 + \frac{1}{m}\right]}.$$

Table 7.11 presents the multiply-imputed results for the leadership data. In this analysis, none of the coefficients are significantly different from zero with $p = 0.05$.

Multiple imputation is now more widely implemented in statistical computing packages. SAS (Yuan 2000) implements multiple imputation procedures with multivariate normal data. The examples in this chapter were computed with Schafer (1999) NORM program, a freeware program for conducting multiple imputation with multivariate normal data. The Appendix provides information about options for obtaining multiple imputation estimates and for combining those estimates.

In general, multiple imputation is the recommended method for handling missing data in any statistical analysis, including meta-analysis. The methods illustrated in this chapter produce divergent results, indicating that the results of this analysis are sensitive to missing data. A potential difficulty in this data could be power since there are just slightly over 40 studies available for analysis. More research is needed to understand the conditions where meta-analysts should use multiple imputation.

Appendix

Computing Packages for Computation of the Multiple Imputation Results

There are a number of options for obtaining multiple imputation results in a meta-analysis model. Two freeware programs are available. The first is the program Norm by Schafer and available at http://www.stat.psu.edu/~jls/misoftwa.html. The Norm program runs as a stand alone program on Windows 95/98/NT. The second is a program available in R by Honaker et al. called Amelia II and available at http://gking.harvard.edu/amelia/. Schafer's norm program was used for the example given earlier.

The program SAS includes two procedures, one for generating the multiple imputations, PROC MI, and a second for analyzing the completed data sets, PROC MIANALYZE. For obtaining the weighted regression results for meta-analysis, the SAS procedure PROC MIANALYZE will have limited utility since the standard errors of the weighted regression coefficients will need to be adjusted as detailed by Lipsey and Wilson (2001). Below is an illustration of the use of PROC MI for the leadership data.

R Programs

One program available in R for generating multiple imputations is Amelia II (Honaker et al. 2011). Directions for using the program are available at http://gking.harvard.edu/amelia/. Once the program is loaded into R, the following command was used to generate $m = 5$ imputed data sets.

> **a.out** < −**amelia(leadimp, m = 5, idvars = "ID")**

Table 7.12 Regression estimates from each imputation generated using Amelia

Variable	Data set 1	Data set 2	Data set 3	Data set 4	Data set 5
Intercept	51.05[a]	68.56	45.79	41.854	−5.974
	(11.28)[b]*	(12.03)*	(11.061)*	(10.984)*	(13.665)
Year	−0.025	−0.034	−0.023	−0.020	0.003
	(0.006)*	(0.006)*	(0.006)*	(0.005)*	(0.007)
Average age	−0.033	−0.019	−0.022	−0.037	−0.019
	(0.004)*	(0.003)*	(0.003)*	(0.004)*	(0.003)*
Percent of male	0.009	0.006	0.006	0.007	0.002
leaders	(0.001)*	(0.001)*	(0.001)*	(0.001)*	(0.001)*
First author	−0.284	−0.127	−0.060	−0.338	−0.154
female	(0.045)*	(0.040)*	(0.033)*	(0.048)*	(0.041)*
Size of	−0.113	−0.056	−0.140	−0.237	−0.001
organization	(0.039)*	(0.037)	(0.042)*	(0.047)*	(0.034)*
Random	0.049	0.101	0.059	0.075	0.068
selection used	(0.019)*	(0.023)*	(0.019)*	(0.020)*	(0.021)*

[a]Coefficient estimate
[b]Standard error of coefficient in parentheses
*Coefficient is significantly different from zero

Table 7.13 Multiply-imputed estimates from Amelia

Variable	Coefficient	SE	Z	p
Intercept	40.255	32.660	1.232	0.217
Year	−0.0198	0.016	−1.207	0.210
Average age	−0.026	0.010	−2.617	0.116
Percent of men	0.006	0.003	1.989	0.148
First author female	−0.193	0.133	−1.451	0.192
Size of organization	−0.110	0.106	−1.039	0.127
Random selection used	0.070	0.030	2.377	0.244

The imputed data sets can be saved for export into another program to complete the analyses using the command,

>**write.amelia(obj = a.out, file.stem = "outdata").**

where "obj" refers to the name given to the object with the imputed data sets (the result of using the command **Amelia**), and "file.stem" provides the name of the data sets that will be written from the program.

Table 7.12 are the weighted regression estimates for the effect size model from each imputation obtained in Amelia. The two variables missing observations are average age of subjects and percent of male leaders. There is variation among the five data sets in their estimates of the regression coefficients. This variation signals that there is some uncertainty in the data set due to missing observations.

Table 7.13 provides the multiply-imputed estimates for the linear model of effect size. These estimates were combined in Excel, and are fairly consistent with the earlier multiple imputation analysis using Schafer's Norm program. None of the coefficients are significantly different from zero.

SAS Proc MI

The SAS procedure PROC MI provides a number of options for analyzing data with missing data. For the example illustrated in this chapter, we use the Monte Carlo Markov Chain with a single chain for the multiple imputations. We also use the EM estimates as the initial starting values for the MCMC analysis. The commands below were used with the leadership data to produce the five imputed data sets:

```
proc mi data = work.leader out = work.leaderimp seed = 101897;
var year ageave perlead gen2 sizeorg2 rndm2 effsize;
mcmc;
```

The first line of the command gives the name of the data set to use, the name of the created SAS data set with the imputations, and the seed number for the pseudo-random number generator. The second command line provides the variables to use in the imputations. Note that the effect size is included in this analysis. The third line specifies the use of Markov Chain Monte Carlo to obtain the estimates of the joint posterior distribution as described by Rubin (1987). Note that the number of imputations are not specified; the default number of imputed data sets generated is five, the number recommended by Schafer (1997).

SAS Proc MI provides a number of useful tables, including one outlining the missing data patterns and the group means for each variable within each missing data pattern. Once the imputations are generated, the procedure gives the estimates for the mean and standard error of the variables with missing data as illustrated below.

Multiple Imputation Parameter Estimates

Variable	Mean	SE	95% confidence limits		DF
Average age of sample	44.109	1.619	40.341	47.877	7.596
Percent of male leaders	65.691	2.898	59.743	71.640	26.869

Variable	Minimum	Maximum	Mu0	t for Mean = Mu0	Pr > |t|
Average age of sample	42.659	45.481	0	27.25	<.0001
Percent of male leaders	64.586	67.390	0	22.67	<.0001

To obtain the weighted regression results for each imputation, we use Proc Reg with weights. The command lines are shown below.

Table 7.14 Multiple imputations generated using SAS Proc MI

Variable	Data set 1	Data set 2	Data set 3	Data set 4	Data set 5
Intercept	29.36[a]	23.70	−7.962	71.548	14.935
	(14.38)[b]*	(11.49)*	(14.16)	(12.560)*	(11.71)
Year	−0.012	−0.012	0.004	−0.036	−0.007
	(0.006)*	(0.006)*	(0.007)	(0.006)*	(0.006)
Average age	−0.026	−0.026	−0.019	−0.013	−0.024
	(0.004)*	(0.004)*	(0.004)*	(0.003)*	(0.003)*
Percent of male	0.009	0.009	0.004	0.005	0.004
leaders	(0.001)*	(0.001)*	(0.001)*	(0.001)*	(0.001)*
First author female	−0.198	−0.198	−0.171	−0.053	−0.202
	(0.042)*	(0.042)*	(0.042)*	(0.037)	(0.040)*
Size of organization	−0.038	−0.038	−0.072	−0.084	0.072
	(0.034)	(0.034)	(0.034)*	(0.034)*	(0.041)*
Random selection	0.006	0.062	0.066	−0.021	0.034
used	(0.031)*	(0.031)*	(0.033)*	(0.028)*	(0.029)

[a]Coefficient estimate
[b]Standard error of coefficient in parentheses
*Coefficient is significantly different from zero

Table 7.15 Multiply-imputed estimates generated by SAS

Variable	Coefficient	SE	t	p
Intercept	26.315	34.305	0.767	0.292
Year	−0.013	0.017	−0.749	0.295
Average age	−0.020	0.007	−2.744	0.111
Percent of men	0.005	0.003	1.706	0.169
First author female	−0.143	0.084	−1.705	0.169
Size of organization	−0.026	0.078	−0.336	0.397
Random selection used	0.030	0.050	0.591	0.330

```
proc reg data = work.leaderimp outest = work.regout covout;
model effsize = year ageave perlead gen2 sizeorg2 rndm2;
weight wt;
by _Imputation_;
run;
```

The lines given above use the SAS data set generated by Proc MI, and estimate the coefficients for the effect size model using weighted regression. The results are computed for each imputation as indicated in the **by** statement. Table 7.14 provides the weighted regression results for each imputation.

Table 7.15 gives the multiply-imputed estimates for the weighted regression results. As in the prior analyses, none of the regression coefficients were significantly different from zero.

References

Allison, P.D. 2002. *Missing data*. Thousand Oaks: Sage.

Begg, C.B., and J.A. Berlin. 1988. Publication bias: A problem in interpreting medical data (with discussion). *Journal of the Royal Statistical Society Series A* 151(2): 419–463.

Buck, S.F. 1960. A method of estimation of missing values in multivariate data suitable for use with an electronic computer. *Journal of the Royal Statistical Society Series B* 22(2): 302–303.

Chan, A.-W., A. Hrobjartsson, M.T. Haahr, P.C. Gotzsche, and D.G. Altman. 2004. Empirical evidence for selective reporting of outcomes in randomized trials. *Journal of the American Medical Association* 291(20): 2457–2465.

Dempster, A.P., N.M. Laird, and D.B. Rubin. 1977. Maximum likelihood from incomplete data via the EM algorithm. *Journal of the Royal Statistical Society Series B* 39(1): 1–38.

Duval, S. 2005. The Trim and Fill method. In *Publication bias in meta-analysis: Prevention, assessment and adjustments*, ed. H.R. Rothstein, A.J. Sutton, and M. Borenstein. West Sussex: Wiley.

Duval, S., and R. Tweedie. 2000. Trim and fill: A simple funnel plot based method of testing and adjusting for publication bias in meta-analysis. *Biometrics* 56(2): 455–463.

Eagly, A.H., M.C. Johannesen-Schmidt, and M.L. van Engen. 2003. Transformational, transactional, and laissez-faire leadership styles: A meta-analysis comparing women and men. *Psychological Bulletin* 129(4): 569–592.

Egger, M., G.D. Smith, M. Schneider, and C. Minder. 1997. Bias in meta-analysis detected by a simple, graphical test. *British Medical Journal* 315(7109): 629–634.

Enders, C.K. 2010. *Applied missing data analysis. Methodology in the Social Sciences*. New York: Guilford.

Fahrbach, K.R. 2001. *An investigation of methods for mixed-model meta-analysis in the presence of missing data*. Lansing: Michigan State University.

Glasser, M. 1964. Linear regression analysis with missing observations among the independent variables. *Journal of the American Statistical Association* 59(307): 834–844.

Hackshaw, A.K., M.R. Law, and N.J. Wald. 1997. The accumulated evidence on lung cancer and environmentaly tobacco smoke. *British Medical Journal* 315(7114): 980–988.

Haitovsky, Y. 1968. Missing data in regression analysis. *Journal of the royal Statistical Society Series B* 30(1): 67–82.

Hemminki, E. 1980. Study of information submitted by drug companies to licensing authorities. *British Medical Journal* 280(6217): 833–836.

Honaker, J., G. King, and M. Blackwell (2011) Amelia II: A program for missing data. http://r.iq. harvard.edu/src/contrib/

Kim, J.-O., and J. Curry. 1977. The treatment of missing data in multivariate analysis. *Sociological Methods and Research* 6(2): 215–240.

Lipsey, M.W., and D.B. Wilson. 2001. *Practical meta-analysis*. Thousand Oaks: Sage Publications.

Little, R.J.A., and D.B. Rubin. 1987. *Statistical analysis with missing data*. New York: Wiley.

Orwin, R.G., and D.S. Cordray. 1985. Effects of deficient reporting on meta-analysis: A conceptual framework and reanalysis. *Psychological Bulletin* 97(1): 134–147.

Rosenthal, R. 1979. The file drawer problem and tolerance for null results. *Psychological Bulletin* 86(3): 638–641.

Rothstein, H.R., A.J. Sutton, and M. Borenstein. 2005. *Publication bias in meta-analysis: Prevention, Assessment and Adjustments*. West Sussex: Wiley.

Rubin, D.B. 1976. Inference and missing data. *Biometrika* 63(3): 581–592.

Rubin, D.B. 1987. Multiple imputation for nonresponse in surveys. Wiley, New York, NY

Schafer, J.L. 1997. *Analysis of incomplete multivariate data*. London: Chapman Hall.

Schafer, J.L. 1999. *NORM: Multiple imputation of incomplete multivariate data under a normal model. Software for Windows*. University Park: Department of Statistics, Penn State University.

Schafer, J.L., and J.W. Graham. 2002. Missing data: Our view of the state of the art. *Psychological Methods* 7(2): 147–177.

Shadish, W.R., L. Robinson, and C. Lu. 1999. *ES: A computer program and manual for effect size calculation*. St. Paul: Assessment Systems Corporation.

Sirin, S.R. 2005. Socioeconomic status and academic achievement: A meta-analytic review of research. *Review of Educational Research* 75(3): 417–453. doi:10.3102/00346543075003417.

Smith, M.L. 1980. Publication bias and meta-analysis. *Evaluation in Education* 4: 22–24.

Sterne, J.A.C., B.J. Becker, and M. Egger. 2005. The funnel plot. In *Publication bias in meta-analysis: Prevention, assessment and adjustment*, ed. H.R. Rothstein, A.J. Sutton, and M. Borenstein. West Sussex: Wiley.

Vevea, J.L., and C.M. Woods. 2005. Publication bias in research synthesis: Sensitivity analysis using a priori weight functions. *Psychological Methods* 10(4): 428–443.

Williamson, P.R., C. Gamble, D.G. Altman, and J.L. Hutton. 2005. Outcome selection biase in meta-analysis. *Statistical Methods in Medical Research* 14(5): 515–524.

Wilson, D.B. 2010. Practical meta-analysis effect size calculator. Campbell Collaboration. http://www.campbellcollaboration.org/resources/effect_size_input.php. Accessed 16 July 2011.

Yuan, Y.C. 2000. Multiple imputation for missing data: Concepts and new developments. http://support.sas.com/rnd/app/papers/multipleimputation.pdf. Accessed 2 April 2011.

Chapter 8
Including Individual Participant Data in Meta-analysis

Abstract This chapter introduces methods for including individual participant data in a traditional meta-analysis. Since meta-analyses use data aggregated to the study, there is potential for aggregation bias, finding relationships between the effect size and study characteristics that may hold only at the level of the study. The potential of aggregation bias may limit the application of meta-analysis results to practice and policy. This chapter provides an example of including publicly available data in a traditional meta-analysis.

8.1 Background

Since Glass (1976) coined the term "meta-analysis," researchers have continued to develop statistical techniques for synthesizing results across studies (Cooper et al. 2009; Hedges and Olkin 1985; Hunter and Schmidt 2004; Rosenthal 1991). In education and psychology, these developments have centered on aggregated data meta-analysis, using summary statistics from primary studies to compute an effect size that is combined across studies. In all meta-analyses, the results depend on the quality and depth of the information provided in a primary study. Cooper and Patall (2009) note that research reviewers are constrained by the information provided by the primary authors, and often face difficulties since authors' reporting practices differ. Researchers conducting syntheses also find that studies fail to report enough statistical data to compute an effect size, or incompletely report information about the study itself for use in moderator analyses. In medicine, meta-analysis researchers have advocated the use of individual patient data as one possible solution to the problem of missing data across studies. Meta-analyses using individual participant data (IPD) have been conducted in a number of reviews, including in colorectal cancer treatments, the effects of anti-depressant drugs in moderately depressed patients (Fournier et al. 2009), and in the study of the relationship between maternal age and type I diabetes (Cardwell et al. 2010).

T.D. Pigott, *Advances in Meta-Analysis*, Statistics for Social and Behavioral Sciences, 109
DOI 10.1007/978-1-4614-2278-5_8, © Springer Science+Business Media, LLC 2012

In medicine, IPD meta-analysis has been referred to as the gold standard of meta-analysis (Simmonds and Higgins 2007). Analyses that use individual level data may lead to conclusions that differ from meta-analyses that use only the aggregated data presented in a study (Schmid et al. 2004) due to problems associated with aggregation and ecological bias. Relationships found between effect size and study methods across studies may not reflect the relationship that exists within studies. One advantage of IPD meta-analysis over traditional or aggregated data meta-analysis (AD) is the increased opportunities to fit nested models of the effects of interventions, recognizing that variation in treatment effects could be due to factors both within and between studies.

This chapter will illustrate methods for including individual participant data in a traditional meta-analysis. The methods discussed in this chapter focus on meta-analyses that include a mix of individual participant data and aggregated study level data, a situation likely to be the most common application of IPD meta-analysis in the social sciences. The example used is the meta-analysis by Sirin (2005) that includes several publicly available data sets that can illustrate how to combine individual participant data with aggregated data from a study into a single meta-analysis.

8.2   The Potential for IPD Meta-analysis

While IPD meta-analysis has been used in medicine, there exists only one example of its use in the social sciences; Goldstein et al. (2000) combined data from studies of the effects of class size with primary data from the Tennessee STAR experiment. IPD appears more in medicine since there is a longer tradition in the medical sciences to register clinical trials and to archive data. The opportunities to analyze pooled data sets in the social sciences are expanding given the increased attention given to data warehousing by both the National Institute of Health (2003) and the National Science Foundation (n.d.). The National Institute of Health (2003) statement on sharing research data indicates that all applications with direct costs above $500,000 must address data sharing. The Social and Economic Science division of the National Science Foundation (n.d.) requires investigators to have a written plan for the archiving of quantitative social and economic data sets collected with NSF support. Journals are also making use of websites for posting computer code or technical appendices that cannot be included in the published article, and may also be a forum for archiving primary study data.

There exist a number of advantages for individual participant data meta-analysis. As mentioned above, individual-level data may alleviate problems with missing data, either for missing effect sizes, or for potential moderators. With the original data, effect sizes can be computed with full information, and analyses of effect size variation can use more detailed background characteristics of the study and participants.

Another potential benefit of IPD meta-analysis is in statistical power. Lambert et al. (2002) compare the power for detecting interactions among study level characteristics and effect sizes in an AD meta-analysis versus an IPD meta-analysis. Under many conditions, an IPD meta-analysis has greater power than an AD meta-analysis. These conditions depend on the within- and between-study variation in the moderators of study effect sizes. When studies have moderators whose values differ widely across studies, then AD meta-analysis using meta-regression will have at least as much power as an IPD meta-analysis, and is much less costly (Cooper and Patall 2009). Alternatively, when moderators vary within studies and are related to the within-study outcome, only an IPD meta-analysis will be able to estimate these within-study interactions. Increased statistical power has also been cited as a reason for encouraging the use of pooled data analysis and data warehousing (Schneider 2010). Many studies of educational interventions use small samples, and pooling these studies could provide more power for investigating the differential effects of these interventions.

As researchers using multi-level modeling have long stressed, aggregation bias operates within nested educational data (Raudenbush and Bryk 2002) and should be carefully monitored in conclusions of AD meta-analysis (Cooper and Patall 2009; Schmid et al. 2004). Having the individual participant data allows the examination of differences in treatment effectiveness at the level of the individual rather than at the level of the study. Being able to make inferences at the individual participant level not only avoids aggregation bias, it may lead to inferences for a meta-analysis that are more readily applied to practice. If we can model how treatment effectiveness varies across studies and also across students within those studies, we have much more information about what works.

A negative example of how AD meta-analysis results can be misinterpreted is the controversy over the use of mammography as a screening tool for breast cancer. While an AD meta-analysis may find that across studies, there is little benefit for screening in women aged 39–49, the meta-analysis cannot tell us about how screening may work for an individual woman with particular characteristics, an analysis that can only be carried out with individual level information about participants. The AD meta-analysis cannot necessarily provide evidence for individual women to make a decision about mammography, only how the benefits of screening vary across studies. More refined analyses using individual student characteristics could lead to inferences that are more directly related to practice than those from data aggregated to the level of the study. One goal of research synthesis is the ability to contribute to decisions about effective treatments. In an IPD meta-analysis, information at the level of student allows stronger inferences without the threat of aggregation bias.

While there are many potential benefits for IPD meta-analysis in educational research, the fact remains that the costs of IPD are much higher than for AD meta-analysis (Shrout 2009; Cooper and Patall 2009). A study estimating the relationship between oral contraceptive use and ovarian cancer (Steinberg et al. 1997) found that the individual participant data meta-analysis on this topic cost approximately five times that of the aggregated meta-analysis. The most costly activities concern the gathering of the data from multiple researchers, and organizing the data into a form that is comparable across studies. Social scientists use of IPD will likely be limited to supplementing aggregated study data with individual level data from data sources that are publicly available. Below I introduce the basic methods of IPD meta-analysis, and illustrate how to use these methods when there is a mix of both aggregated data as in a traditional meta-analysis, and individual-level data.

8.3   The Two-Stage Method for a Mix of IPD and AD

There are two methods for analyzing a mix of individual participant data and aggregated data: the one-stage method, and the two-stage method. The one-stage method is based on multi-level modeling techniques, while the two-stage method aggregates the individual participant data to the study level, obtaining essentially the same information as a reviewer would extract from a primary research study, and then uses standard meta-analysis methods. In order to present the models that will be used, I begin with a brief outline of random effects models for aggregated data meta-analysis, and then discuss the two-stage model for combining both aggregated data and individual-level data.

8.3.1   Simple Random Effects Models with Aggregated Data

In order to present the models in this chapter, I will need to introduce notation for the simple random effects model. Recall from Chap. 2 we refer to our generic effect size as T_i where $i=1,\ldots,k$ with a within-study variance for the effect size denoted by v_i. When we use a random effects model, we assume that the variance of our effect sizes has two components, one due to sampling variance, v_i, and one due to the variance in the underlying population of effect sizes, τ^2. The random effects variance for our effect size is denoted by $v_i^* = v_i + \tau^2$. We can write the random effects model using two levels. With θ_i^* as our underlying population effect size for study i, we can write our estimated effect size, T_i, as in Raudenbush (2009) as

$$T_i = \theta_i^* + e_i, \text{ where } e_i \sim \eta(0, v_i). \tag{8.1}$$

The residual for study i, e_i, is normally distributed with a mean of 0 and a variance of v_i, which is known in AD meta-analysis. Level-2 is the model of the true effect sizes, θ_i^*, given as

$$\theta_i^* = \theta^* + u_i, \text{ where } u_i \sim \eta(0, \tau^2) \tag{8.2}$$

and where θ^* is the overall mean effect size in the population, and the residual u_i is assumed normally distributed with mean 0 and variance, τ^2. Thus the total variance in the observed effect sizes, T_i, is $v_i^* = v_i + \tau^2$. The random effects variance, τ^2, can be estimated either directly using the method of moments, or with restricted maximum likelihood as detailed by Raudenbush (2009) and illustrated later in the chapter when discussing one-stage methods. We then estimate the mean effect size θ^* as the random effects weighted mean of the T_i, or

$$\hat{\theta}^* = \bar{T}_{\bullet}^* = \frac{\sum_{i=1}^{k} T_i/v_i^*}{\sum_{i=1}^{k} 1/v_i^*}. \tag{8.3}$$

with variance given by

$$\mathrm{var}(\hat{\theta}^*) = \frac{1}{\sum_{i=1}^{k} 1/v_i^*}. \tag{8.4}$$

The model above can be used with the standardized mean difference. With correlation coefficients, the AD model is different from the one we normally use in standard meta-analysis practice. Common practice in the synthesis of correlations is to use Fisher's z-transformation to normalize the distribution of the correlations as detailed in many texts on meta-analysis (Borenstein et al. 2009; Lipsey and Wilson 2000). In this paper, we will use the correlation as the effect size without transforming it in order to be consistent with the one-stage models presented later. Thus, we can write our simple random effects model for the correlations, r_i, from each study $i = 1,\ldots,k$, as

$$
\begin{aligned}
r_i &= \theta_i^* + e_i \\
\theta_i^* &= \theta^* + u_i \\
e_i &\sim \eta(0, v_i) \\
u_i &\sim \eta(0, \tau^2)
\end{aligned}
\tag{8.5}
$$

where within studies, the correlation, r_i, estimates the underlying population value, θ_i^*, and e_i is distributed normally with mean 0 and variance v_i which is considered known in the meta-analysis context, and is given below. Each studies' underlying population mean effect size is also considered sampled from a distribution of effects sizes with overall mean equal to θ^* and u_i distributed normally with mean 0 and variance equal to τ^2. Within studies, the variance of the correlation is considered known and is given by

$$v_i = \frac{(1 - r_i^2)^2}{n} \tag{8.6}$$

Hedges and Olkin (1985) note that this approximation to the variance for the correlation is unbiased when the sample size of the study is at least 15, a situation that holds in our example

8.3.2   Two-Stage Estimation with Both Individual Level and Aggregated Data

As Riley et al. (2008) state, the easiest method to employ with a mix of IPD and AD is a two-stage model. The researcher first computes the study level effect sizes from each IPD study, and then continues with estimating the random effects model given in (8.1) for the standardized mean difference, and in (8.5) for the correlation coefficient. For example, if we are using the standardized mean difference between the treatment and control group as our effect size, we would use the individual level data from study i to compute the treatment and control group means, $\bar{Y}_{trt}$, and $\bar{Y}_{cntl}$, and the pooled standard deviation of the outcome, s_p. We would then obtain the effect size, T_i, for study i, and estimate the model given in (8.1). Chapter 3 provides computing options for estimating the variance component, τ^2.

8.3.2.1   Example: Two-Stage Method Using Correlation as the Effect Size

Sirin (2005) reports on a meta-analysis of studies that estimate the association between socioeconomic status and academic achievement. Included in the Sirin (2005) meta-analysis is a number of publicly-available data sets, including the National Educational Longitudinal Survey (NELS), the National Longitudinal Study of Youth (NLSY), the Longitudinal Study of American Youth, and the National Transition Demonstration Project. These data sets can be used at the individual-participant level to obtain an estimate of the correlation between measures of socioeconomic status and achievement. The table in the Data Appendix provides the correlations between SES and achievement for the NELS and the NLSY data set that will be used for the two-stage analysis.

For the two-stage method, we take the correlations estimated from NELS and NLSY along with those from the aggregated data in the Data Appendix and compute a simple random effects model. Note that we are using the correlation in this model rather than Fisher's z-transformation. Using SAS Proc Mixed (the code is given in the Appendix) yields an estimate of the random effects mean correlation of $\hat{\theta}^* = 0.283$ with an estimate of the variance component, $\tau^2 = 0.0336$. The histograms of the correlations and their Fisher-z transformations both appear fairly normal, and thus the analysis is assumed robust to the distribution of the correlations in this data. The variance component and mean correlation in this example are computed using restricted maximum likelihood (REML). The output from SAS Proc Mixed gives both the random effects mean and the variance component in one step, unlike the method of moments discussed in Chap. 3. Recall that in the method of moments, we first compute τ^2, add this value to our fixed effects study weights, and then compute the mean effect size with the newly constructed weights. Using REML in SAS, we can do the estimation of τ^2 and θ^* in one step.

Though the two-stage method is easily computed, it suffers from the same issues as all aggregated data analysis – only relationships at the level of the study are estimated leaving open the possibility of aggregation bias. Reviewers must exercise caution in applying results from a meta-analysis using only aggregated study-level data to within-study relationships. For example, while we may find that an intervention effect is not homogeneous across studies, this result does not imply that the intervention effect is not homogeneous among participants within a study.

8.4   The One-Stage Method for a Mix of IPD and AD

In order to illustrate the one-stage method, we begin with a model for the data from a study that provides only individual participant data. The model with only individual participant data will use the individual level measures, with a parameter in the model that represents the target effect size. I will illustrate the model for IPD with both the standardized mean difference and the correlation.

8.4.1   IPD Model for the Standardized Mean Difference

The standardized mean difference is used when we have a study that utilizes two experimental groups, say a treatment group and a control group. The outcome for the IPD model is the individual participant's response on the target measurement, denoted for participant j, in study i, denoted as y_{ij}. In order to make the outcomes parallel in an IPD analysis with an AD analysis using the standardized mean difference, we will use the standardized outcome, denoted here by Z_{Yij} for student j, $j = 1,\ldots, n_j$, in study i, $i=1,\ldots,k$. Thus, each students' outcome will be standardized using the overall mean and standard deviation of the outcome observed in that study. We can write a hierarchical linear or mixed model for our IPD data following Riley et al. (2008). For a study that uses the standardized mean difference as the effect size, the model can be given as

$$
\begin{aligned}
Z_{Yij} &= \phi_i + \theta_i^* \, x_{ij} + e_{ij} \\
\theta_i^* &= \theta^* + u_i \\
e_{ij} &\sim \eta(0,\ 1) \\
u_i &\sim \eta(0,\ \tau^2)
\end{aligned}
\tag{8.7}
$$

where x_{ij} is a 0/1 code designating control or treatment group membership, the fixed study effect is ϕ_i, with the random treatment effect in study i given by θ_i^*. Given the parameters in (8.7), the value for θ_i^* will be the difference between the treatment and control group, standardized by the pooled sample standard deviation which is equivalent to the standardized mean difference in an AD meta-analysis, T_i. The variance within each study for the outcome is 1, since we have standardized our outcomes, and the variance for the θ_i^* is τ^2. Our goal in an IPD meta-analysis would be to estimate the mean treatment effect, θ^*, and its variance, τ^2, using standard methods of hierarchical linear models (Raudenbush and Bryk 2002). The individual level models rely on parameterizing the variables in IPD so that one of the regression coefficients will be equal to the estimate of the effect size in a meta-analysis that uses aggregated data.

8.4.2   IPD Model for the Correlation

As in the model for IPD with the standardized mean difference, we can write a model for our correlation effect sizes that includes the target correlation as a coefficient in the model. If we are interested in the correlation between Y and X, using the standardized version of X to predict the standardized version of Y results in the regression coefficient equal to the correlation between the two variables. Between studies, we remain interested in the estimate of the random effect mean correlation and its random effects variance. Using the standardized versions of our variables Y_{ij} and X_{ij} as given by Z_{Yij} and Z_{Xij}, our model can be written as

$$
\begin{aligned}
Z_{Yij} &= \phi_i + \theta_i^* Z_{Xij} + e_{ij} \\
\theta_i^* &= \theta^* + u_i \\
e_{ij} &\sim \eta(0, 1) \\
u_i &\sim \eta(0, \tau^2)
\end{aligned}
\tag{8.8}
$$

where ϕ_i is the fixed effect for study i, θ_i^* is the target correlation for the meta-analysis, and e_{ij} is distributed normally with mean 0 and variance 1 since we standardized the variables. Note that $i=1,\ldots,k$ where k is the number of studies, and $j=1,\ldots,n_i$ where n_i is the number of cases in study i. The second line of (8.8) is the same as in (8.5), where the correlation from each study is assumed to estimate a grand mean, θ^*, with u_i distributed normally with mean 0 and variance equal to τ^2.

8.4.3   Model for the One-Stage Method with Both IPD and AD

One-stage methods for a mix of IPD and AD would be analogous to fitting hierarchical linear models when some of the level-2 units do not provide individual level data.

When we combine both types of data using the one-stage method, the AD studies will contribute only the effect size and its variance, while the studies with IPD will contribute individual level data. For the standardized mean difference this will be the standardized outcome, and for correlations, the standardized versions of the two variables whose correlation is the review's focus.

We will use the models given in (8.7) and (8.8) for both AD and IPD, but the values that the IPD studies and the AD studies contribute to the model will be different. For example, the AD studies will have a single value for Z_{Yij} when we are interested in the standardized mean difference. For correlations, the value for Z_{Xij} will be equal to 1 and Z_{Yij} will equal the study correlation, r_i, so that (8.8) simplifies to the level 1 model given in (8.5) for the AD studies.

To combine AD and IPD in the one-stage method, we will add a dummy variable, D_i, that takes the value 1 when the study contributes individual level data, and 0 when the study contributes aggregated data. For the standardized mean difference, the model given by Riley et al. (2008) is

$$
\begin{aligned}
Z'_{Yij} &= D_i\,\phi_i + \theta_i^* x_{ij} + e'_{ij} \\
\theta_i^* &= \theta^* + u_i \\
e'_{ij} &\sim \eta(0,\ v'_i) \\
u_i &\sim \eta(0,\ \tau^2)
\end{aligned}
\tag{8.9}
$$

For each IPD study, the outcome is Z_{Yij}, the standardized value of the outcome, and the $v'_i = 1$ since we have standardized our outcomes within studies. The dummy code $D_i = 1$, and ϕ_i is the fixed study effect. In the AD studies, we assume only one observation ($j = 1$), and we set $x_{ij} = 1$, and $D_i = 0$. The response in the AD studies is the estimate of the effect size in that study, $Z'_{Yij} = T_i$, and $v'_{ij} = v_i$, the within-study sampling variance of T_i.

When the outcome is the correlation coefficient, the model is given by

$$
\begin{aligned}
Z'_{Yij} &= D_i\,\phi_i + \theta_i^* Z'_{Xij} + e'_{ij} \\
\theta_i^* &= \theta^* + u_i \\
e'_{ij} &\sim \eta(0,\ v'_i) \\
u_i &\sim \eta(0,\ \tau^2)
\end{aligned}
\tag{8.10}
$$

For the IPD studies, the outcome is $Z'_{Yij} = Z_{Yij}$ and $Z'_{Xij} = Z_{Xij}$, the standardized versions of the variables in the target correlation. As in the standardized mean difference case, the $v'_i = 1$ since the variables are standardized. The dummy code $D_i = 1$, and ϕ_i is the fixed study effect. In the AD studies, we also assume only one observation per study, and we set our outcome equal to $Z'_{Yij} = r_i$. We also set $D_i = 0$, and $Z'_{Xij} = 1$ so that for the AD studies, (8.10) reduces to (8.5), with v_i defined in (8.6).

8.4.3.1   Example: One-Stage Method with Correlations

In the Sirin data, we are interested in estimating the correlation between measures of student achievement and measures of socio-economic status. The estimated parameter θ^* is then the correlation of interest in the meta-analysis. The one-stage model for correlations can be written as

$$Z'ach_{ij} = D_i\,\phi_i + \theta_i^*\,Z'ses_{ij} + e'_{ij}$$
$$\theta_i^* = \theta^* + u_i$$
$$e'_{ij} \sim \eta(0,\,v'_i)$$
$$u_i \sim \eta(0,\,\tau^2)$$

(8.11)

Following the model given by Riley et al., the dummy code D_i takes the value 1 when the study provides individual participant data, and 0 when the study provides only aggregated data. For each IPD study, the outcome, $Z'ach_{ij}$, is the standardized value for the achievement test in study i, $Z'ses_{ij}$ is the standardized value for the measure of SES in study i, and thus, $v'_i = 1$ since we have standardized the variables. When both variables are standardized, then the θ_i^* estimated in the first equation above is equal to the correlation between the two variables. In the AD studies, we assume only one observation so that $j = 1$, $Z'ach_{ij} = r_i$, $Z'ses_{ij} = 1$, and the variance is approximated by $v'_i = v_i$ given in (8.6). This formulation for the AD studies then simplifies to the equation for the one-way random effects model.

Using SAS Proc Mixed (the code is given in the Appendix), the one-step method provides an estimate of the random effects mean correlation, $\hat{\theta}^* = 0.312$, with an estimate of the variance component, $\tau^2 = 0.03404$. Both estimates are slightly larger than that for the two-step method.

8.5   Effect Size Models with Moderators Using a Mix of IPD and AD

One advantage of IPD emphasized in the literature (Cooper and Patall 2009; Higgins et al. 2001; Riley et al. 2008) is the ability to fit effect size moderator models that have more power than meta-regression models with only aggregated data. Simmonds and Higgins (2007) demonstate that the power of meta-regression in aggregated data meta-analysis as compared to individual level data depends on the variation in the target moderator. If the effects of treatment vary within studies as a function of the moderator, then an IPD analysis will have more power. However, if the moderator values vary across studies, then the AD meta-analysis will have more power to detect these study-level relationships. With a mix of IPD and AD, the models discussed in this section can examine the relationships between potential moderators and effect size both within and between studies. For example, we may see that within a study, age may not be related to the effectiveness of a treatment, but across studies, due to the variation in samples, age does appear related to effect size magnitude. Below I present the two-stage method for meta-regression models with a mix of IPD and AD, and then the one-stage method.

8.5.1   Two-Stage Methods for Meta-regression with a Mix of IPD and AD

As in the two-stage method for computing the mean effect size, the two-stage method for a meta-regression involves aggregating the individual level data to the level of the study, and then proceeding with standard meta-regression techniques. For a two-stage analysis using both IPD and AD, and with the standardized mean difference as our effect size, we would fit a random effects meta-regression given by

$$
\begin{aligned}
T_i &= \theta_i^* + \beta \bar{m}_{i\bullet} + e_i \\
\theta_i^* &= \theta^* + u_i \\
e_i &\sim \eta(0, v_i) \\
u_i &\sim \eta(0, \tau^2)
\end{aligned}
\tag{8.12}
$$

The mean value of the moderator variable (for example, mean grade level) is denoted by $\bar{m}_{i\bullet}$ for study i. This random effects meta-regression model is a simple bivariate model examining the relationship between a single moderator and the effect size. In a standard meta-regression application, β is the weighted least-squares regression coefficient for the predictor m. For studies with IPD, we would estimate the effect size T_i and the mean of the moderator $\bar{m}_{i\bullet}$ using the individual level data, and then add those values to the aggregated data set to estimate the meta-regression model.

When the effect sizes in our studies are correlations, we have a model similar to that in (8.12). We have an estimate of the correlation, r_i, and the variance of that correlation, v_i, given in (8.6). We also have a moderator variable, $\bar{m}_{i\bullet}$, or the study level mean of a moderator, m, such as the average age of the study participants, or the percentage of minority students included in the sample. We can write the random effects meta-regression as

$$
\begin{aligned}
r_i &= \theta_i^* + \beta \bar{m}_{i\bullet} + e_i \\
\theta_i^* &= \theta^* + u_i \\
e_i &\sim \eta(0, v_i) \\
u_i &\sim \eta(0, \tau^2)
\end{aligned}
\tag{8.13}
$$

Table 8.1 Estimates from meta-regression using two-step method

Effect	Estimate	SE	Lower CI	Upper CI
θ^*	0.356	0.050	0.258	0.454
% minority	−0.146	0.082	−0.306	0.014

As in the model for the standardized mean differences, with IPD, we would compute the target correlation, and the study level mean for the moderator, and add these values to the aggregated data.

8.5.1.1   Example: Two-Stage Model with a Mix of IPD and AD

One of the potential moderators in the Sirin (2005) meta-analysis of the correlation between socio-economic status and academic achievement is the percent of minority students in the study sample. In the NELS and NLSY studies, we have the racial background for each individual, and thus can aggregate to the study level the percentage of minority (non-white) students in each of these two data sets. The data table in the Data Appendix gives the aggregated percent of minority students for the sample of NELS and NLSY participants in this meta-analysis, and the percent of minority students in the AD studies. Using SAS Proc Mixed (code given in the Appendix), we find in Table 8.1 that the percentage of minority students included in the study was inversely related to the study's correlation, however this effect was not statistically significant, (β = -0.146, S.E. = 0.082, C.I.: -0.306, 0.014). The estimate of the random effects mean correlation, θ^*, was 0.356 (with S.E. = 0.050, 95% C.I.: 0.258, 0.454), slightly larger than in the simple random effects analysis. The estimate of the random effects variance was slightly smaller than in the simple mixed effects model ($\tau^2 = 0.0321$). The small decrease in the variance estimate is due to the inclusion of the across-study moderator, percent minority students in the sample.

As in the example in Sect. 8.3.2.1, we are using REML as our estimation procedure for the variance component. SAS Proc Mixed allows the estimation of $\hat{\theta}^*$, $\hat{\beta}$ and τ^2 in one step.

8.5.2   One-Stage Method for Meta-regression with a Mix of IPD and AD

As in the discussion of the simple random effects model for IPD and AD, I will first present a model that assumes only individual level data. Then we will combine that model with the one given in 8.12 and (8.13). If we only had individual level data, we would be adding a moderator to the model given in (8.7) and (8.8). For the standardized mean difference, the model in (8.7) uses the standardized outcome, Z_{Yij}, and a dummy code, x_{ij}, to indicate membership in the treatment or control group. For correlations, the model in (8.8) has the standardized version of one of our variables, Z_{Yij}, as the outcome and one predictor, the standardized version of the other variable, Z_{Xij}.

8.5.3 Meta-regression for IPD Data Only

I first present the model for the standardized mean difference when we only have IPD. Let m_{ij} be the value for the moderator variable centered at the mean for the study i, person j. This value could be for example, the actual age of the participant, or a dummy code for whether the individual is a minority or not, centered at the study mean value for the predictor. The IPD model, with interactions between the predictors, is given by

$$
\begin{aligned}
Z_{Yij} &= \phi_i + \theta_i^* x_{ij} + \gamma_1 m_{ij} + \gamma_2 x_{ij} m_{ij} + \beta x_{ij} \bar{m}_{i\bullet} + e_{ij} \\
\theta_i^* &= \theta^* + u_i \\
e_{ij} &\sim \eta(0, v_i) \\
u_i &\sim \eta(0, \tau^2)
\end{aligned} \tag{8.14}
$$

In this model, we have interaction effects with the moderator at both the within-study and between-study levels. The two coefficients, γ_1 and γ_2, capture the within-study effect of the moderator m_{ij} and its interaction with x_{ij}, respectively. At the between-study level, we have the interaction effect between x_{ij} and the study-level mean of the moderator, $\bar{m}_{i\bullet}$, for study i.

The model for studies with only IPD and correlations as the effect size is similar to (8.14), and is given by

$$
\begin{aligned}
Z_{Yij} &= \phi_i + \theta_i^* Z_{Xij} + \gamma_1 m_{ij} + \gamma_2 Z_{Xij} m_{ij} + \beta Z_{Xij} \bar{m}_{i\bullet} + e_{ij} \\
\theta_i^* &= \theta^* + u_i \\
e_{ij} &\sim \eta(0, v_i) \\
u_i &\sim \eta(0, \tau^2)
\end{aligned} \tag{8.15}
$$

As in the model in (8.14), we estimate two coefficients within the study, one for the effect for the moderator, m_{ij}, and the second for the interaction between Z_{Xij} and the moderator, m_{ij}. At the between-study level, we have the interaction effect between Z_{Xij} and the study-level mean of the moderator, $\bar{m}_{i\bullet}$, for study i.

8.5.4 One-Stage Meta-regression with a Mix of IPD and AD

To add studies with aggregated data to the model in (8.14) and (8.15), we need to use the dummy code, D_i, which takes the value 1 when the study contributes individual level data and 0 when the study contributes aggregated data. For the standardized mean difference, the model is

$$
\begin{aligned}
Z'_{Yij} &= D_i \phi_i + \theta_i^* x_{ij} + D_i \gamma_1 m_{ij} + D_i \gamma_2 x_{ij} m_{ij} + \beta x_{ij} \bar{m}_{i\bullet} + e'_{ij} \\
\theta_i^* &= \theta^* + u_i \\
e'_{ij} &\sim \eta(0, v'_i) \\
u_i &\sim \eta(0, \tau^2)
\end{aligned} \tag{8.16}
$$

As in (8.9), the outcome for IPD studies is the standardized value of y_{ij}, or $Z'_{Yij} = Z_{Yij}$, and $v'_i = 1$. For AD studies, we assume only one observation for that study, and set the outcome as $Z'_{Yij} = T_i$, the standardized mean difference for that study. The other values are set as $x_{ij} = 1$, the $D_i = 0$, and $v'_i = v_i$, the within study sampling variance for the effect size, T_i.

For the correlation coefficient, the model is

$$Z'_{Yij} = D_i \, \phi_i + \theta_i^* \, Z'_{Xij} + D_i \gamma_1 \, m_{ij} + D_i \gamma_2 \, Z'_{Xij} \, m_{ij} + \beta \, Z'_{Xij} \, \bar{m}_{i\bullet} + e'_{ij}$$
$$\theta_i^* = \theta^* + u_i$$
$$e'_{ij} \sim \eta(0, v'_i)$$
$$u_i \sim \eta(0, \tau^2)$$

$$(8.17)$$

The values for IPD studies are the same as in (8.10), where $Z'_{Yij} = Z_{Yij}$ and $Z'_{Xij} = Z_{Xij}$ are the standardized versions of the variables in the target correlation. The value of the variance is $v'_i = 1$ since the variables are all standardized. In the AD studies, we only have one observation where $Z'_{Yij} = r_i$ and $Z'_{Xij} = 1$. The variance $v'_i = v_i$, the within-study sampling variance of the r_i. Since $D_i = 0$, (8.17) simplifies to (8.13) when the study contributes only aggregated data.

8.5.4.1   Example: One-Stage Method for Meta-regression with Correlations

As in Example 8.5.1.1, an important study-level moderator in the Sirin (2005) data is the percent of minority students in the sample. The magnitude of the correlation between achievement and socioeconomic status may vary with race. Within the IPD studies, we can code each observation using a dummy code m_{ij} which takes the value 0 if the student is identified as White, and 1 if the student is either Asian, Black, Hispanic or Native American. We will use the centered version of the dummy code in the analysis to be consistent with the AD model at the study level. At the study level, the mean of these dummy codes, $\bar{m}_{i\bullet}$, corresponds to the percent of minority students in the sample. For the one-step method, the combined model for the Sirin data can be given as

$$Z' ach_{ij} = D_i \, \phi_i + \theta_i^* \, Z' ses_{ij} + D_i \gamma_1 \, m_{ij} + D_i \gamma_2 \, Z' ses_{ij} \, m_{ij}$$
$$+ \beta \, Z' ses_{ij} \, \bar{m}_{i\bullet} + e'_{ij}$$
$$\theta_i^* = \theta^* + u_i$$
$$e'_{ij} \sim \eta(0, v'_i)$$
$$u_i \sim \eta(0, \tau^2)$$

$$(8.18)$$

Using SAS Proc Mixed (the code is given in the Appendix) to fit the model in (8.18) across both AD and IPD studies, we obtain the following:

Effect	Estimate	SE	Lower CI	Upper CI
θ^*	0.362	0.022	0.319	0.406
γ_1 (minority)	−0.200	0.018	−0.235	−0.166
γ_2 (interaction of minority and SES)	−0.089	0.016	−0.122	−0.056
β (% minority in study)	−0.167	0.058	−0.280	−0.054

The variance component for this analysis is $\tau^2 = 0.0328$. Note that in this analysis, the between-studies moderator, percent minority students in the sample, is significantly related to the correlation, as indicated by the estimate for β. We have more power in this analysis given the individual participant data to estimate this relationship than in the two-step method. Across studies, the correlation between achievement and SES is smaller for studies with a larger percentage of minority students. Within the NELS and NLSY studies, minority students, in general, have a lower correlation between SES and achievement, as indicated by the estimate for γ_1. There is also a significant interaction between minority and SES. As seen in the value for γ_2, minority students with higher SES have an even lower correlation between SES and achievement than White students with a similar SES. This interaction is not apparent unless we have individual level data.

The procedures used here can be applied beyond the context of traditional meta-analysis. The current federal education policies are moving toward state-based accountability systems. In order to compare differences across states, we will need methodology that can combine data across states that use different types of assessments, where some of the individual-level data will be available, while other data will be aggregated to the state or district. The methods of IPD meta-analysis could provide more sensitive analyses that take into account the nested nature of the data. As in the Sirin (2005) example above, there could be relationships between variables within studies that cannot be examined with a typical AD meta-analysis.

Appendix: SAS Code for Meta-analyses Using a Mix of IPD and AD

SAS Code for Simple Random Effects Model Using the Two-Step Method

Below is the code needed to run a random effects meta-analysis with the raw correlation. The model for this analysis is given in (8.5). The data for this analysis should include a variable "outcome" that holds the effect size for each study (in this case, the correlation coefficient), a variable that provides a unique identifier for each study, labeled "study" in this example. We also need the variances for each effect size estimate, in this example, the estimated variance of the correlation coefficient as given in (8.6). The first line of code calls the SAS procedure Proc Mixed. The options in this line, **noclprint** and **covtest** are options for output. The **noclprint** suppresses the printing of class level information, or the list of all studies in this example. The **covtest** prints out the standard errors and test statistics for the variance and covariance parameters. The **class** statement indicates that "study" is a class variable, which will be designated as the random effect later in the code. The **model** statement asks for a simple random effects model with "outcome" as the dependent variable. The option **solution** prints out the fixed-effects parameter estimates, which in this example is the estimate of the mean effect size ("outcome"). The **random** statement designates the random effect, and the option **solution** prints the estimate of the random effect, or in this case, the random effects variance. The **repeated** statement specifies the covariance matrix for the error terms, and in this case, allows between group heterogeneity. **The parms** statement provides the starting values for the covariance parameters. The first value is the starting value for the overall variance component for this model. The next 42 elements are the within-study estimates of the variance of the effect size, here the correlation. The option **eqcons** fixes the variances for the 42 studies in the analysis since these are considered known in a meta-analysis model.

```
proc mixed noclprint covtest;
class study;
model outcome =/solution;
random study/solution;
repeated/group = study;
parms
(0.05)
(.000071) (.000311) (.001584) (.008978) (.000222) (.000041) (.000072)
(.004656) (.002213) (.007193) (.002426) (.008054) (.002803) (.003252)
(.000616) (.005791) (.000557) (.001512) (.000676) (.001345) (.002302)
(.002746) (.003254) (.007465) (.010415) (.004995) (.008849) (.002823)
(.002765) (.000083) (.000221) (.002669) (.000665) (.002237) (.000598)
(.000924) (.000813) (.001663) (.003136) (.001659) (.027218) (.005051)
/eqcons=2 to 43;
run;
```

Output from Two-Stage Simple Random Effects Model

Table 8.2 gives the estimate of the variance component in the first line along with its standard error and test of significance. Table 8.3 provides the random effects mean for the overall effect size.

Table 8.2 Estimate of variance component from SAS Proc Mixed

Covariance parameter estimates

Cov parm	Group	Estimate	Standard error	Z value	Pr > Z
Study		0.03362	0.007880	4.27	< .0001
Residual	Study 1Nels	0.000071	0	.	.
Residual	Study 2Nlsy	0.000311	0	.	.
Residual	Study alexent	0.001584	0	.	.
. . . . . .	. . . . . .	. . . .			
Residual	Study watk	0.005051	0	.	.

Table 8.3 Estimate of random effects mean from SAS Proc Mixed

Solution for fixed effects

Effect	Estimate	Standard error	DF	t value	Pr > |t|
Study	0.2830	0.02954	41	9.58	< .0001

SAS Code for Meta-regression Using the Two-Stage Method

The code here is the same as for the aggregated data simple random effects model except for the addition of a moderator in the **model** statement. The model for this analysis is given in (8.8). The moderator in this example is **permin**, the percent of minority students in the study sample. As in the example above, restricted maximum likelihood is the default estimation method.

```
proc mixed noclprint covtest;
class study;
model outcome = permin/solution;
random study/solution;
repeated/group = study;
parms
(0.05)
(.000071) (.000311) (.001584) (.008978) (.000222) (.000041) (.000072)
(.004656) (.002213) (.007193) (.002426) (.008054) (.002803) (.003252)
(.000616) (.005791) (.000557) (.001512) (.000676) (.001345) (.002302)
(.002746) (.003254) (.007465) (.010415) (.004995) (.008849) (.002823)
(.002765) (.000083) (.000221) (.002669) (.000665) (.002237) (.000598)
(.000924) (.000813) (.001663) (.003136) (.001659) (.027218) (.005051)
/eqcons=2 to 43;
run;
```

Table 8.4 Estimate of variance component from SAS Proc Mixed for meta-regression

Covariance parameter estimates					
Cov parm	Group	Estimate	Standard error	Z value	Pr > Z
Study		0.03207	0.007574	4.23	< .0001
Residual	Study 1Nels	0.000071	0	.	.
Residual	Study 2Nlsy	0.000311	0	.	.
Residual	Study alexent	0.001584	0	.	.
....	....	....			
Residual	Study watk	0.005051	0	.	.

Table 8.5 Estimates for meta-regression using SAS Proc Mixed

Solution for fixed effects					
Effect	Estimate	Standard error	DF	t value	Pr > |t|
Intercept	0.3560	0.05016	40	7.10	< .0001
Permin	−0.1456	0.08170	0	−1.78	.

Output from Meta-regression Using the Two-Stage Method

Table 8.4 gives the estimate of the variance component in the first line along with its standard error and test of significance conditional on the moderator **permin**. Table 8.5 provides the random effects mean for the overall effect size given the percent of minority students in the sample. Note that the slope for **permin** is not statistically significant, and thus in this analysis, the percent of the minority students in the sample is not associated with the correlation between SES and achievement in these studies.

Table 8.6 Example of data for a mixed IPD and AD analysis

Study	Person	ipd	Zses	Outcome
Watk	1	0	1	0.3600
Whitr	1	0	1	0.1540
NELS	0001	1	1.38817	1.90799
NLSY	0803	1	1.38817	0.87512

SAS Code for Simple Random Effects Model Using the One-Stage Model

The code below estimates a simple random effects model with both IPD and AD data. The data is set up as in Table 8.6. Two of the AD studies are given in the first two lines of the table. The AD studies contribute one observation (Person =1), does not provide individual level data (IPD = 0), has a value of the standardized SES of

1, and has an outcome equal to the correlation estimated in that study. The last lines provide examples of data from individuals in the NELS and NLSY data set. Each individual has a person identification number, has a value of IPD=1 since individual level data is provided, and also provides a standardized value for SES and for the outcome which is achievement in our example.

Given the values for the variables in Table 8.6, the SAS code follows the general structure of the examples for the AD random effects models. The first line calls the SAS Proc Mixed, with option **cl** and **noclprint** defined as in earlier sections. The option **method = reml** specifies the estimation method as restricted maximum likelihood (the default). Since we have two levels in our model – the within-study level for the studies that provide IPD, and the between-study level, we have two class variables, person and study. The **model** statement reflects the model discussed in (8.13). The outcome for the IPD studies, achievement in its standardized form, will be modeled with study as a factor, and **zses** as a predictor. Since ipd =0 for the AD studies and zses = 1, the model for the AD studies will be equivalent to the model given for simple random effects as given above. The options for the model command are **noint** which fits a no-intercept model, **s** which is short for **solution** and provides the solution for the fixed effects parameters. The option **cl** produces the confidence limits for the covariance parameter estimates, and **covb** gives the estimated variance-covariance matrix of the fixed-effects parameters. The **random** command line indicates that study is a random effect. The command line for **repeated** specifies the variance-covariance matrix for the mixed model. The option **type = un** specifies that the variance-covariance matrix is unstructured. The option **subject = study (person)** indicates that the variable persons is nested within study. The option **group=study** indicates that observations having the same value of **study** are at the same level and should have the same covariance parameters. The **parms** command provides a number of values for the variance parameters. The first three values after **parms** are the starting values for the random effects variances, the first for the overall random effects variance between studies, the second two for the NELS and the NLSY studies. The next set of parameters is the within-study variances for the outcome or effect size.

```
proc mixed noclprint cl method=reml;
class person study;
model outcome = study*ipd zses/noint s cl covb;
random study;
repeated/type = un subject = study(person) group = study;
parms
(0.05) (0.05) (0.05)
(0.001584129) (0.008978216) (0.000222147) (4.14666E-05)
(7.23005E-05) (0.004655734) (0.002213119) (0.007192762)
(0.000221388) (0.002668852) (0.000664976) (0.002236663)
(0.002425768) (0.008054047) (0.002802574) (0.00325176)
(0.000616328) (0.005790909) (0.000557348) (0.000597936)
(0.000923595) (0.001511538) (0.000676163) (0.001345186)
```

(0.002302425) (0.002746439) (0.003253767) (0.000812903)
(0.007464759) (0.010414583) (0.004995399) (0.008849115)
(0.002822828) (0.002765381) (8.29676E-05) (0.001659)
(0.001662966) (0.003136358) (0.02721837) (0.005050641)
/eqcons= 4 to 43;
run;

Table 8.7 Estimate of variance component from SAS Proc Mixed for simple random effects model using the one-step method

Cov parm	Subject	Group	Estimate	Alpha	Lower	Upper
Study			0.03404	0.05	0.0223	0.0583
UN(1,1)	Study(person)	Study 1Nels	0.8937	0.05	0.8709	0.9176
UN(1,1)	Study(person)	Study 2Nlsy	0.9324	0.05	0.8853	0.9835
UN(1,1)	Study(person)	Study alexent	0.001584	.	.	.
....		....	....			
UN(1,1)	Study(person)	Study watk	0.005051	.	.	.

Table 8.8 Estimates for simple random effects model using SAS Proc Mixed

Effect	Study	Estimate	Standard error	DF	t value	Pr > |t|
ipd*study	Nels	0.000	0.1847	1403	0.00	1.000
ipd*study	Nlsy	−0.00052	0.1854	1403	0.00	0.9978
ipd*study	Alexent	0	.	.	.	.
...	...	..	..	..	..	..
ipd*study	Watk	0	..	..	..	..
Zses		0.3125	0.007769	1403	40.22	<.0001

Output from One-Stage Simple Random Effects Model

The important parts of this output are under Covariance Parameter Estimates, which gives the estimate of the variance component for the simple random effects model with both individual-level and study-level data. The table below looks different from the one in the simple random effects model with only study-level data since one option used here was **cl**, the confidence limits for the covariance parameters. The second table gives the estimate of the random effects mean effect size, as the solution for the fixed effects for the variable *Zses* (Tables 8.7 and 8.8).

SAS Code for a Meta-regression Model Using the One-Step Method

The SAS code for a meta-regression moderator model in the one-step method includes the same code as for the simple random effects models as given above. The main difference occurs in the "model" statement, where we add a number of variables and interactions to the model. The "ipd*min" variable is the within-study effect of being a minority student on the association between achievement and SES within the IPD studies. The dummy code for minority student has also been centered at the study mean. The factor "idp*zses*min" is the interaction effect between SES and being a minority student on student achievement. The "zses*permin" effect is the between-study association of the percent of minority students with the study-level correlation between achievement and SES. The rest of the command lines are the same as in the one-step method for the simple random effects model.

```
proc mixed cl noclprint method=reml;
class person study;
model outcome = newid*ipd zses ipd*min ipd*zses*min zses*permin/noint s cl
covb;
random study;
repeated/type = un subject = study(person) group = study;
parms
(0.05) (0.05) (0.05)
(0.001584129) (0.008978216) (0.000222147) (4.14666E-05)
(7.23005E-05) (0.004655734) (0.002213119) (0.007192762)
(0.000221388) (0.002668852) (0.000664976) (0.002236663)
(0.002425768) (0.008054047) (0.002802574) (0.00325176)
(0.000616328) (0.005790909) (0.000557348) (0.000597936)
(0.000923595) (0.001511538) (0.000676163) (0.001345186)
(0.002302425) (0.002746439) (0.003253767) (0.000812903)
(0.007464759) (0.010414583) (0.004995399) (0.008849115)
(0.002822828) (0.002765381) (8.29676E-05) (0.001659)
(0.001662966) (0.003136358) (0.02721837) (0.005050641)
/eqcons= 4 to 43;
run;
```

Output for Meta-regression Using the One-Step Method

Table 8.9 below gives the conditional random variance component for the model, 0.0319, which is similar to the meta-regression estimate using the two-step method. In Table 8.10, we see the estimates for the multilevel model for the one-step method. The overall estimate of the random effects conditional mean is 0.362, similar to the two-step result, with the overall estimate of the between-study effect of percent of

Table 8.9 Estimate of the variance component from SAS Proc Mixed for meta-regression model using one-step method

Covariance parameter estimates

Cov parm	Subject	Group	Estimate	Alpha	Lower	Upper
Study			0.03284	0.05	0.0217	0.0556
UN(1,1)	Study(person)	Study 1Nels	0.9015	0.05	0.8781	0.9258
UN(1,1)	Study(person)	Study 2Nlsy	0.8558	0.05	0.8111	0.9044
UN(1,1)	Study(person)	Study alexent	0.001584	.	.	.
....		....	....			
UN(1,1)	Study(person)	Study watk	0.005051	.	.	.

Table 8.10 Estimates for meta-regression in one-step method

Solution for fixed effects

Effect	Study	Estimate	Standard error	DF	t value	Pr > |t|
ipd*study	Nels	−0.0089	0.1814	1403	−0.05	0.9607
ipd*study	Nlsy	0.0001	0.1821	1403	0.00	0.999
ipd*study	Alexent	0	.	.	.	.
...	...	..	..	..	..	..
ipd*study	Watk	0	..	..	..	..
Zses		0.3622	0.02205	1403	16.43	<.0001
idp*min		−0.2007	0.01754	1403	−11.44	<.0001
idp*zses*min		−0.08902	0.01678	1403	−5.30	<.0001
Zses*permin		−0.1674	0.05755	1403	−2.91	0.0036

minority students as −0.167, which in this analysis is statistically significant. The two within-study factors, being a minority study, and the interaction between minority status and SES, are also both significant as seen in the lines just before the bottom of the second table.

References

Borenstein, M., L.V. Hedges, J.P.T. Higgins, and H.R. Rothstein. 2009. *Introduction to meta-analysis*. Chicester: Wiley.

Cardwell, C.R., L.C. Stene, G. Joner, M.K. Bulsara, O. Cinek, J. Rosenbauer, and J. Ludvigsson. 2010. Maternal age at birth and childhood Type 1 diabetes: A pooled analysis of observational studies. *Diabetes* 59: 486–494.

Cooper, H., L.V. Hedges, and J.C. Valentine (eds.). 2009. *The handbook of research synthesis and meta-analysis*. New York: Russell Sage.

Cooper, H., and E.A. Patall. 2009. The relative benefits of meta-analysis conducted with individual participant data versus aggregated data. *Psychological Methods* 14: 165–176.

Fournier, J.C., R.J. DeRubeis, S.D. Hollon, S. Dimidjian, J.D. Amsterdam, R.D. Shelton, and J. Fawcett. 2009. Antidepressant drug effects and depresssion severity: A patient-level meta-analysis. *Journal of the American Medical Association* 303(1): 47–53.

Glass, G.V. 1976. Primary, secondary, and meta-analysis. *Educational Researcher* 5(10): 3–8.

Goldstein, H., M. Yang, R. Omar, R. Turner, and S. Thompson. 2000. Meta-analysis using multilevel models with an application to the study of class size effects. *Applied Statistics* 49: 399–412.

Hedges, L.V., and I. Olkin. 1985. *Statistical methods for meta-analysis*. New York: Academic.

Higgins, J.P.T., A. Whitehead, R.M. Turner, R.Z. Omar, and S.G. Thompson. 2001. Meta-analysis of continuous outcome data from individual patients. *Statistics in Medicine* 20: 2219–2241.

Hunter, J.E., and F.L. Schmidt. 2004. *Methods of meta-analysis: Correcting error and bias in research findings*, 2nd ed. Thousand Oaks: Sage.

Lambert, P.C., A.J. Sutton, K.R. Abrams, and D.R. Jones. 2002. A comparison of summary patient-level covariates in meta-regression with individual patient data meta-analysis. *Journal of Clinical Epidemiology* 55: 86–94.

Lipsey, M.W., and D.B. Wilson. 2000. *Practical meta-analysis*. Thousand Oaks: Sage.

National Institutes of Health. 2003. Final NIH statement on sharing research data. NIH.

National Science Foundation Directorate for Social BES. n.d. Data archiving policy. National Science Foundation. http://www.nsf.gov/sbe/ses/common/archive.jsp. Accessed 29 May 2010.

Raudenbush, S.R. 2009. Analyzing effect sizes: Random-effects models. In *The handbook of research synthesis and meta-analysis*, 2nd ed, ed. H. Cooper, L.V. Hedges, and J.C. Valentine. New York: Russell Sage.

Raudenbush, S.R., and A.S. Bryk. 2002. *Hierarchical linear models: Applications and data analysis methods*, 2nd ed. Thousand Oaks: Sage.

Riley, R.D., P.C. Lambert, J.A. Staessen, J. Wang, F. Gueyffier, L. Thijs, and F. Bourtitie. 2008. Meta-analysis of continuous outcomes combining individual patient data and aggregate data. *Statistics in Medicine* 27: 1870–1893.

Rosenthal, R. 1991. *Meta-analytic procedures for social research*, 2nd ed. Newbury Park: Sage.

Schmid, C.H., P.C. Stark, J.A. Berlin, P. Landais, and J. Lau. 2004. Meta-regression detected associations between heterogeneous treatment effects and study-level, but not patient-level, factors. *Journal of Clinical Epidemiology* 57: 683–697.

Schneider, B. 2010. Personal communication, March 2010.

Shrout, P.E. 2009. Short and long views of integrative data analysis: Comments on contributions to the special issue. *Psychological Methods* 14(2): 177–181.

Simmonds, M.C., and J.P.T. Higgins. 2007. Covariate heterogeneity in meta-analysis: Criteria for deciding between meta-regression and individual patient data. *Statistics in Medicine* 26: 2982–2999.

Sirin, S.R. 2005. Socioeconomic status and academic achievement: A meta-analytic review of research. *Review of Educational Research* 75(3): 417–453. doi:10.3102/00346543075003417.

Steinberg, K.K., S.J. Smith, D.F. Stoup, I. Olkin, N.C. Lee, G.D. Williamson, and S.B. Thacker. 1997. Comparison of effect estimates from a meta-analysis of summary data from published studies and from a meta-analysis using individual patient data for ovarian cancer studies. *American Journal of Epidemiology* 145: 917–925.

Chapter 9
Generalizations from Meta-analysis

Abstract This chapter discusses the kinds of inferences and generalizations we can make from a meta-analysis. The chapter reviews the framework outlined by Shadish et al. (2002) for meta-analysis, and provides examples from two recent syntheses that had an influence on policy.

9.1 Background

What kinds of decisions can we make from a meta-analysis? Are we justified in making policy decisions from the results of a meta-analysis about implementation of an intervention, such as the use of systematic phonics instruction? What about decisions of a more personal nature – such as should I have a mammography annually in my forties? These are questions asked about meta-analyses from policy makers, practitioners, and consumers of this information. This chapter reviews the basis whereby we can make inferences from a meta-analysis, and the kinds of inferences that can be supported. It also argues for the transparency that reviewers of evidence should provide so that the results of systematic reviews can be used appropriately.

As many researchers have pointed out, the results of meta-analyses are observational. Reviewers cannot manipulate the kinds of methods used, or the participants in the sample, and thus cannot fulfill the requirements of an experimental study that aims to identify causes. In a meta-analysis, we do not have the ability to assign the conditions of the study. The studies already exist, and use a variety of procedures, methods, participants, in a variety of settings. Despite this fact, we still want to make decisions about the types of interventions that are most effective, or about what personal choice I should make about my own health. Given the nature of meta-analysis, we cannot use statistical reasoning as we do in a randomized controlled trial to reach a causal inference. Instead, we need a different basis for arguing about cause.

T.D. Pigott, *Advances in Meta-Analysis*, Statistics for Social and Behavioral Sciences, DOI 10.1007/978-1-4614-2278-5_9, © Springer Science+Business Media, LLC 2012

As Matt and Cook (2009) and Shadish et al. (2002) point out, causal inferences can be supported in a meta-analysis using a logic or basis different from classic arguments about cause. These researchers argue that the warrant for making causal claims from a meta-analysis depends on ruling out threats to the validity of that inference. In other words, we argue a causal claim from a meta-analysis by systematically addressing all other plausible explanations for the causal relationship we propose.

Shadish et al. outline how we can proceed by arguing from the five principles of generalized causal inference: (1) surface similarity, (2) ruling out irrelevancies, (3) making discriminations, (4) interpolation and extrapolation, and (5) causal explanation. Causal inferences from a meta-analysis require examining the conditions and methods used across studies. As Shadish et al. argue, the number of methods and conditions represented across studies in a meta-analysis allows us to have more evidence about how an intervention or a relationship varies than in a single study that cannot include all of these conditions.

This chapter will use the Preventive Health Services (Nelson et al. 2009) report on breast cancer screening and Ehri et al. (2001) meta-analysis on the effects of systematic phonics instruction to illustrate the five principles of generalized causal inference described by Shadish et al. In the discussion that follows, I use Cronbach (1982) acronym, UTOS, as a shorthand for the generalizations we want to make from the meta-analysis. UTOS stands for the Units (persons) who receive the treatment and to whom we wish to generalize, the Treatments in the study and those treatments we want to generalize about, the Observations (measures) used in the study and those we wish to generalize to, and the Settings where the study takes place, and settings where we want to generalize these findings. Below is an introduction to the report on breast cancer screening as well as to the work of the National Reading Panel (2000) followed by a discussion of each of the five principles of generalized causal inference in the context of these reviews.

9.1.1   The Preventive Health Services (2009) Report on Breast Cancer Screening

In 2009, the United States Preventive Health Services (USPHS) released an update of their previous synthesis on studies focusing on the outcomes of breast cancer screening. The update included two new trials, the Age trial from the United Kingdom (Moss et al. 2006), and an update of the data from the Gothenburg trial conducted in Sweden (Bjurstam et al. 2003). The Age trial specifically targeted outcomes in women aged 40–49, and resulted in the USPHS revising their original (US Preventive Services Task Force 2002) recommendations for this specific group of women. Essentially, the recommendations were that the evidence no longer supported routine, annual mammography for women aged 40–49, given the risk of false positive results and over diagnosis. The release of the results coincided with the healthcare reform debates that were occupying the US Congress and the media.

In some instances, the report's recommendations were linked to the healthcare reform debate (Woolf 2010), one clear example of how the results of this review were misinterpreted. There have been many commentaries in the media and in the medical literature discussing what conclusions can be drawn, and how women should respond to the report.

9.1.2 The National Reading Panel's Meta-analysis on Learning to Read

The U. S. Congress in 1997 asked that a panel be convened to review research on strategies to teach children to read. The panel, appointed by the National Institute of Child Health and Human Development (NICHD) and the Department of Education conducted a series of systematic reviews of the evidence. One of these subgroup analyses is a meta-analysis conducted by Ehri et al. (2001) on the effects of systematic phonics instruction on students' ability to read words. The National Panel's report had wide-spread influence, contributing to the language in the No Child Left Behind Act (Allington 2006) that calls for the use of research-based teaching strategies. The National Panel's report was widely criticized on various grounds (Camilli et al. 2006; Hammill and Swanson 2006; Pressley et al. 2004), with many researchers questioning how well the results could generalize to real classrooms. Below I outline the five principles of generalized causal inference, using both the Ehri et al. meta-analysis and the breast cancer screening study as examples.

9.2 Principles of Generalized Causal Inference

9.2.1 Surface Similarity

Surface similarity was first discussed by Campbell (1957) in terms of construct validity. We can more safely apply a generalization from one measure to another measure that is based on a similar construct. In the context of meta-analysis, we can apply a finding from a meta-analysis to those UTOS that are represented in the studies included in the synthesis. Conversely, we may caution about generalizing from a meta-analysis to a context that is not represented in the meta-analysis. One criticism of the breast cancer screening meta-analysis was that the studies in the meta-analysis did not include a sufficient number of African-American women, and thus the results should not be generalized to this group of women. For example, Murphy (2010) cautions that clinicians applying the findings of the report need to keep in mind that African-American women have a higher risk of mortality from breast cancer, and women of Ashkenazi Jewish descent are at higher risk of genetically mediated breast cancer. The synthesis did not include studies with a large sample of these groups of women, and thus the synthesis provides no evidence about how breast cancer screening is related to the mortality from breast cancer for these two groups. In terms of other groups of women who might be at higher risk, such as women exposed to high levels of radiation, there is not enough specific information included about the characteristics of women in the trial to make generalizations about particular sub-groups. Our ability to examine surface similarity from the report itself is limited. The report does update findings about one particular group of women, those aged 40–49, since new evidence from the Age Trial (Moss et al. 2006) provides more direct evidence about this group. There were, however, no new trials that could provide insight for the screening of women over the age of 70, and thus the report does not revise the guidelines for women in this age group.

In the debate over the National Reading Panel's meta-analysis on systematic phonics, Garan (2001) questioned the use of measures of different reading outcomes as equivalent in the meta-analysis. The Ehri et al. meta-analysis used the construct of general literacy to include decoding regular words, decoding pseudowords, spelling words and reading text orally to name a few. To Garan, these measures are not sufficiently similar to each other to constitute a single construct. In the Ehri et al. review, the effect sizes for the different measures are reported separately though they are treated as measuring the effectiveness of programs on systematic phonics.

9.2.2   Ruling Out Irrelevancies

Related to surface similarity is the principle of ruling out irrelevancies. In order to generalize a finding to a set of UTOS that were not represented in the meta-analysis, we need to understand whether a given situation is similar to the ones represented in the meta-analysis, and what differences between our given situation and those in the meta-analysis are irrelevant to the findings. In the breast cancer screening review, one issue deemed irrelevant to mortality of breast screening is whether the mammography used film or digital technology. The research question guiding the review includes both of these mammography procedures, but does not provide a comparison of their effectiveness on mortality outcomes in the review. Thus, the reviewers conducted the review on the assumption that film and digital mammography lead to the same mortality rates. However, some researchers do raise issues about whether the method of screening is really irrelevant. For example, Berg (2010) presents evidence that magnetic resonance imaging (MRI) for women at high risk improves detection by 40% over mammography and ultrasound combined. The report does not compare outcomes using MRI versus film or digital technology. Murphy (2010) also suggests that to avoid higher rates of false positives, younger women should consider having their screening at facilities with radiologists that focus on breast imaging and that use digital technology. Here Murphy questions whether film versus digital technology is actually an irrelevant factor. It may not be possible to test empirically whether outcomes are different between film and digital mammography with the current evidence, so this may be an area that needs more research.

Camilli et al. (2006), in their review of the findings of the National Reading Panel on systematic phonics, note that the meta-analysis compares treatments that received various levels of systematic phonics with a no-treatment control. Camilli et al. argues that while the report's findings (as indicated in the title the Ehri et al. 2001) states that systematic phonics increases student reading achievement, the meta-analysis itself did not and could not examine the differences among the different types of systematic phonics programs represented in the sample of studies. Thus, we are not able to determine from this meta-analysis whether the difference among systematic phonics programs in the amount of phonics instruction is an irrelevant factor.

9.2.3 Making Discriminations

As Shadish et al. (2002) describe, we make discriminations in a meta-analysis about the conditions where the cause and effect relationship does not hold, or, in other words, for those persons, treatments, measures and settings where the findings are found not to apply. This principle is different from surface similarities in that it refers to the examination of moderators of a given cause and effect relationship. For example, Littell et al. (2005) has found that the reported effectiveness of multisystemic therapy for at-risk children varies as a function of the involvement of the researcher in the development of the intervention. Studies that were conducted by researchers other than the original developers have smaller effect sizes. We can think of this finding as discriminating about the conditions where the treatment is most effective. The breast cancer screening study is limited in its ability to make discriminations partly due to the lack of information about the backgrounds of the women involved in the studies, and partly due to the small number of trials (seven). Moderator analyses examining how the results might vary systematically among persons, treatments, measures and settings are not possible since there are only seven trials that meet the inclusion criterion. We do not have enough statistical power to make discriminations about the relative effectiveness of screening across different UTOS.

One finding from the Ehri et al. (2001) meta-analysis that was not subject to debate was that systematic phonics instruction did not appear as effective for older elementary school children as for those in kindergarten and first grade. This finding was based on a number of studies that included older children. In fact, Ehri et al. used simple moderator analyses to examine both grade and reading ability, finding that kindergartners and first graders at risk had the largest benefit from systematic phonics instruction. Children in 2nd through 6th grade had little benefit.

9.2.4   *Interpolation and Extrapolation*

Another interrelated principle is interpolation and extrapolation. In examining a causal claim from a meta-analysis, we need to specify the range of characteristics of UTOS where the cause and effect relationship applies. In a single, primary study, we are careful not to extrapolate to contexts outside of the ones represented in the study itself – a single study cannot provide much evidence about whether the findings hold outside of the UTOS used in the study. In some meta-analyses, we could have a wide range of persons, treatments, measures and settings represented across studies, and we can systematically examine whether the cause and effect relationship applies across the studies. One method for interpolating and extrapolating studies is to use modeling strategies with effect sizes, using meta-regression, for example, to see what combinations of characteristics of studies may find larger or smaller effect sizes. As described above, the breast cancer screening review does not include enough studies to model the range of possible study characteristics where the results do or do not apply. The breast cancer screening review does not make recommendations on the effects of screening on women older than 70 – the sample of studies simply does not provide evidence about this group, and the authors of the review do not extrapolate the results. One classic example of the use of modeling in this way is illustrated in Raudenbush and Bryk (1985). Using a random effects meta-regression model, Raudenbush shows that the effect size in the teacher expectancy studies drops off considerably when the induction of expectancy is performed after the teachers have known their students for 3 weeks or more.

One issue of extrapolation and interpolation raised in the systematic phonics meta-analysis relates to the nature of the phonics programs. As Camilli et al. (2006) explains, the reading treatment described in the literature can rarely be classified as including systematic phonics instruction versus less systematic phonics instruction as might occur in a classroom where phonics is only taught when needed. Underlying this criticism of the phonics meta-analysis is the question of whether the phonics treatment as described in these studies is implemented in a similar way in classrooms. Pearson (2004) raises this question in a history of the whole language movement prior to the National Reading Panel report; teachers were less involved and invested in the critiques around the National Reading Panel and No Child Left Behind than academics. The realities of systematic instruction of phonics in a classroom may not resemble the studies in the meta-analysis, and may also be difficult to classify.

9.2.5   *Causal Explanation*

The fifth principle is causal explanation. Though a meta-analysis may not include information about how an intervention works, Shadish et al. (2002) argue that with good theory, meta-analyses can contribute to our understanding about causes.

Causal explanation can be facilitated in a meta-analysis by breaking down the intervention reviewed into its component parts, and positing a theory about both the critical ingredients of an intervention and how those ingredients relate to one another. A meta-analysis can then focus on the parts of this theory of action, using effect size modeling to examine what components of the intervention are most strongly associated with the magnitude of the effect size. In addition, a logic model or theory of action can provide a map of what evidence exists in the literature about particular aspects of a mediating process, and where more studies are needed to provide insight into aspects of the model. In the breast cancer screening study, there are not enough studies to map out an elaborated logic model. However, there are areas in the social sciences that may have the potential of supporting this type of analysis.

Pressley et al. (2004) raise the issue of theory of action or model of reading that is implied by the National Reading Panel work. For Pressley et al., the Reading Panel focused on a set of skills that are related to reading but may be much more narrow than intended. Pressley et al. argue that the theory of reading underlying the National Panel work suggests that "beginning reading only requires instruction in phonemic awareness, phonics, fluency, vocabulary, and comprehension strategies" (p. 41). The criticism of the report may have been tied to this difference in theory of how reading develops in children. The report may not have emphasized enough that the meta-analyses examined component parts of an effective reading program, and were not intended to define a comprehensive reading curriculum.

9.3   Suggestions for Generalizing from a Meta-analysis

Both the breast cancer screening study and the meta-analysis on systematic phonics instruction captured much attention due to the characterization of their findings by various groups. In the breast cancer screening case, the findings appeared to contradict current practice (yearly mammography) particularly in women aged 39–50. The meta-analysis on systematic phonics stirred controversy since its findings were influential on subsequent education policy. The question is what can reviewers do to decrease potential for misinterpreting meta-analysis findings and misapplying them to policy and practice? One suggestion is based on the Cochrane Handbook's (Higgins and Green 2011) risk of bias tables. With the assistance of experts in the field of study, reviewers might attempt a summary of what aspects of UTOS in a given field appear to have enough evidence to make a recommendation, and where we have equivocal or no evidence. Table 9.1 below is an attempt at a table for the Ehri et al. (2001) work.

Table 9.1 is not complete, but may serve as a way to summarize where we do have evidence to take an action. For each element of UTOS, I indicate the level of evidence for particular generalizations from Ehri et al. Both Pearson (2004) and Pressley et al. (2004) mention the role of policymakers in using the results of the National Panel report in ways that went beyond the data gathered. We do need ways

Table 9.1 Outline of generalizations supported in Ehri et al. (2001)

Area	Evidence	Equivocal evidence	No evidence
Units	Adequate for K-1 graders at risk		Second language learners
Treatments		Differences in effectiveness among types of programs, and how much systematic phonics instruction is necessary	
Observations/ measures	Word reading and pseudo –word reading	General reading ability not well defined so that not clear that tests of comprehension are equivalent to tests of work reading	
Settings		No differences found among instructional delivery units of tutoring, small group or whole class	

to communicate complex findings to those who may use our reviews. Those who are interested in this book are by nature interested in meta-analysis and summarizing the evidence in an area, and thus we also must be as careful in how we describe what actually can be done with our results.

References

Allington, R.L. 2006. Reading lessons and federal policymaking: An overview and introduction to the special issue. *The Elementary School Journal* 107: 3–15.

Berg, W.A. 2010. Benefits of screening mammography. *Journal of the American Medical Association* 303(2): 168–169.

Bjurstam, N., L. Bjorneld, J. Warwick, et al. 2003. The Gothenburg breast screening trial. *Cancer* 97(10): 2387–2396.

Camilli, G., P.M. Wolfe, and M.L. Smith. 2006. Meta-analysis and reading policy: Perspectives on teaching children to read. *The Elementary School Journal* 107: 27–36.

Campbell, D.T. 1957. Factors relevant to the validity of experiments in social settings. *Psychological Bulletin* 54(4): 297–312.

Cronbach, L.J. 1982. *Designing evaluations of educational and social programs*. San Francisco: Jossey-Bass.

Ehri, L.C., S. Nunes, S. Stahl, and D. Willows. 2001. Systematic phonics instruction helps students learn to read: Evidence from the National Reading Panel's meta-analysis. *Review of Educational Research* 71: 393–448.

Garan, E.M. 2001. Beyond the smoke and mirrors: A critique of the National Reading Panel report on phonics. *Phi Delta Kappan* 87(7): 500–506.

Hammill, D.D., and H.L. Swanson. 2006. The National Reading Panel's meta-analysis of phonics instruction: Another point of view. *The Elementary School Journal* 107: 17–26.

Higgins, J.P.T., and S. Green. 2011. *Cochrane handbook for systematic reviews of interventions*. Oxford, UK: The Cochrane Collaboration.

Littell J.H., M. Campbell, S. Green, and B. Toews. 2005. Multisystemic therapy for social, emotional and behavioral problems in youth aged 10–17. *Cochrane Database of Systematic Reviews* (4). doi:10.1002/14651858.CD004797.pub4.

Matt, G.E., and T.D. Cook. 2009. Threats to the validity of generalized inferences. In *The handbook of research synthesis and meta-analysis*, ed. H. Cooper, L.V. Hedges, and J.C. Valentine, 537–560. New York: Russell Sage.

Moss, S.M., H. Cuckle, A. Evans, et al. 2006. Effect of mammographic screening from age 40 years on breast cancer mortality at 10 years' follow-up: A randomised controlled trial. *Lancet* 386(9552): 2053–2060.

Murphy, A.M. 2010. Mammography screening for breast cancer: A view from 2 worlds. *Journal of the American Medical Association* 303(2): 166–167.

National Reading Panel. 2000. *Report of the National Reading Panel: Teaching chidren to read: An evidence-based assessment of the scientific research literature on reading and its implications for reading instruction: Reports of the subgroups.* Rockvill: NICHD Clearinghouse.

Nelson, H.D., K. Tyne, A. Naik, C. Bougatsos, B. Chan, P. Nygren, and L. Humphrey. 2009. Screening for breast cancer: Systematic evidence review update for the U. S. Preventive Services Task Force (trans: Agency for Healthcare Research and Quality). Rockville, MD: U. S. Department of Health and Human Services.

Pearson, P.D. 2004. The reading wars. *Educational Policy* 18: 216–252.

Pressley, M., N.K. Duke, and E.C. Boling. 2004. The educational science and scientifically based instruction we need: Lessons from reading research and policymaking. *Harvard Educational Review* 74: 30–61.

Raudenbush, S.W., and A.S. Bryk. 1985. Empirical Bayes meta-analysis. *Journal of Educational Statistics* 10: 75–98.

Shadish, W.R., T.D. Cook, and D.T. Campbell. 2002. *Experimental and quasi-experimental designs for generalized causal inference.* Boston: Houghton Mifflin Company.

US Preventive Services Task Force. 2002. Screening for breast cancer: Recommendations and rationale. *Annals of Internal Medicine* 137(5 Part 1): 344–346.

Woolf, S.H. 2010. The 2009 breast cancer screening recommendations of the US Preventive Services Task Force. *Journal of the American Medical Association* 303(2): 162–163.

Chapter 10
Recommendations for Producing a High Quality Meta-analysis

Abstract This chapter provides a set of recommendations based on prior chapters in the book for improving the quality of meta-analyses.

10.1 Background

The prior chapters of the book illustrate methods for advanced meta-analysis, with the goal of increasing the quality of both the meta-analytic techniques used and the inferences drawn from these reviews. As a summary, this chapter provides recommendations for increasing the quality of the meta-analyses that are produced to inform evidence-based decisions. Systematic reviews that include a meta-analysis represent one consideration used by policymakers to make decisions; as Gibbs (2003) states, preferences of the clients, values of the organization, and the resources available all enter into policy debates for good reason. What I hope is that when a systematic review and meta-analysis can bring evidence to bear on a problem that the review itself fairly represents the literature available, and the data itself. Below I provide recommendations for raising the quality of the meta-analysis part of a systematic review.

10.2 Understanding the Research Problem

A systematic review and meta-analysis requires a deep substantive understanding of the focus area of the review. Completing a research synthesis and meta-analysis requires patience, and many small decisions about how to handle particular studies and data within those studies. Though not every problem can be anticipated prior to conducting a systematic review and meta-analysis, a substantive understanding of the area of research can serve as a guide for making important decisions about the meta-analysis. For example, Chaps. 4, 5, and 6 provide examples of how to

T.D. Pigott, *Advances in Meta-Analysis*, Statistics for Social and Behavioral Sciences, DOI 10.1007/978-1-4614-2278-5_10, © Springer Science+Business Media, LLC 2012

compute the power of the statistical tests in a meta-analysis. A researcher cannot compute power without knowledge of what constitutes a substantively important effect, the typical sample sizes of studies in the area, and the likely number of studies that may exist for synthesis. Another reason for having substantive knowledge of an area appears in Chap. 3. Reviewers need to make decisions about the use of random versus fixed effects models based on the nature of the focus intervention, and/or the characteristics of the studies. Interventions that include multiple components, are difficult to implement with fidelity, or that are widely used may include variability that is likely due to unknown between-study differences and thus is more realistically modeled with random effects.

Another advantage of substantive expertise is the opportunity to include individual participant data. As discussed in Chap. 8, IPD meta-analysis allows analyses of associations within studies, and can add more specific understanding of how interventions, for example, are related to participant characteristics rather than average features of a study sample. A substantive expert may know which publicly available data sets have been used in the area, and may also have informal contacts for obtaining individual-level data from a primary author.

Having deep knowledge of the research problem will also increase the likelihood that the results of the meta-analysis are discussed in ways that contribute to policy and practice. If the reviewer knows the major controversies in the literature either about differential effectiveness of an intervention or about how constructs relate to each other, then the reviewer can direct attention in the research synthesis and meta-analysis toward those issues. If the literature base does not support analyses directed at these issues, then the reviewer is contributing just by pointing out a major gap in the knowledge base. One problem that might not be alleviated by substantive expertise is over-generalizing from meta-analytic results. As Chap. 9 discusses, researchers cannot make causal inferences from meta-analyses in the same way as they can from well-controlled randomized experiments. The process for making inferences from a meta-analysis about possible causal relationships has to be based on ruling out possible reasons for the association found between study characteristics and study results. However, substantive experts may have more background for examining alternative explanations for associations found in a meta-analysis than those new to a field.

10.3   Having an a Priori Plan for the Meta-analysis

With substantive expertise, the reviewers can also develop an a priori plan for the meta-analysis. Creating a logic model or a map of how constructs relate to one another identifies the potential moderators for the analysis. Having a plan can also help avoid the problem of Type I errors when reviewers conduct a series of statistical tests. Reviewers should have an a priori idea of what analyses will be critical, and how to minimize the number of statistical tests included. With an a priori plan, a reviewer can also conduct power analyses of the most substantively important tests to see how many studies would be needed to detect a given effect. Once the literature search and coding are complete, then the reviewer will have a clearer idea of what tests are possible, and what tests will also provide adequate power.

An a priori plan identifying potential moderators will also help reviewers handle missing data when it occurs. If particular moderators are likely to be missing, then reviewers can make sure that other data is coded from a study that could serve as proxies for those moderators, or could be used in a multiple imputation to help model the complete data distribution as illustrated in Chap. 7.

10.4   Carefully and Thoroughly Interpret the Results of Meta-analysis

An understanding of the research problem and an a priori analysis plan should lead a reviewer to a more thorough interpretation of meta-analytic results. Substantive expertise can highlight the important controversies in the literature that can then serve as a basis for an analysis plan. The plan can also alert reviewers to areas where power to test a given question might be inadequate. Identifying the critical issues, and carrying out analyses to address those issues then allows the reviewer to examine carefully the nature of the evidence that can apply to a given problem.

In the interpretation of results, reviewers with substantive knowledge should also focus on ways to transform effect sizes to metrics that readers can understand. Often standardized mean differences can be translated to a metric familiar to readers such as points on a common standardized test like the SAT. Odds ratios can also be discussed in ways that emphasize the different risks between two groups. Not knowing how to interpret effect sizes in substantively useful ways limits readers' understanding and subsequent application of the review's results.

The identified research issues and a priori plan may help researchers being overwhelmed by the amount of data collected in the meta-analysis, and also allow reviewers to be guided by theory rather than the data. An example of this problem is meta-analyses that include a series of one-way ANOVA models. The fact that study results vary on the basis of a single variable at a time does not lead to a coherent conclusion. I find myself wanting to know about combinations of these potential moderators, i.e., whether instruction programs are more effective with all low-income children regardless of age, or whether their effectiveness depends on both income level and grade level. These analyses are possible with meta-regression, and even if meta-regression is not feasible due to limited numbers of studies, exploring how these moderators are related to one another would add to our understanding of what programs work for what types of students.

If the reviewer makes an informed and careful interpretation of the results, then we may also decrease the potential of misinterpretation of the meta-analysis. Presenting tables of one-way ANOVA results increases the likelihood that a reader will make a causal inference based on a single one-way ANOVA, and apply that inference to a policy decision.

Thinking carefully about Shadish et al. (2002) principles of generalized causal inference will also provide a check on the types of inferences made from a meta-analysis. Even if the systematic review yields equivocal results, a well-conducted systematic review should help to illuminate the issues. Using tables such as the one at the end of Chap. 9 to summarize the limits of the inferences possible could lead to better studies in the future, or at least acknowledgement that an evidence-based decision may not be possible given the state of the literature.

More careful and more nuanced interpretation, while not necessarily palatable to policy-makers, may, in fact, increase the usefulness of meta-analyses by providing an accurate picture of how the effectiveness of interventions can vary. In this way, reviewers may help policy-makers to avoid non-evidence-based decision-making, a goal all systematic reviews share.

References

Gibbs, L.E. 2003. *Evidence-based practice for the helping professions: A practical guide with integrated multimedia*. Pacific Grove: Brooks/Cole-Thomson.

Shadish, W.R., T.D. Cook, and D.T. Campbell. 2002. *Experimental and quasi-experimental designs for generalized causal inference*. Boston: Houghton Mifflin Company.

Chapter 11
Data Appendix

11.1 Sirin (2005) Meta-analysis on the Association Between Measures of Socioeconomic Status and Academic Achievement

Sirin (2005) conducted a systematic review of studies reporting a correlation between socioeconomic status (SES) and academic achievement. A number of different measures have been used in the literature for both SES and achievement; the goal of the meta-analysis was to examine whether variation in the strength of the association between SES and achievement varies depending on the types of measures used, and characteristics of the studies and their samples. The data used to construct Table 3.1 through 3.6 are given in Table 11.1 below.

The data below are the cases from Sirin (2005) used in the meta-regression in Chap. 3 (Table 11.2).

T.D. Pigott, *Advances in Meta-Analysis*, Statistics for Social and Behavioral Sciences, DOI 10.1007/978-1-4614-2278-5_11, © Springer Science+Business Media, LLC 2012

Table 11.1 Selected cases from Sirin (2005)

Case	N	r	Achievement measure	SES measure
1	453	0.391	GPA	Free lunch
2	39	0.719	State Test	Free lunch
3	106	0.072	State Test	Free lunch
4	85	0.467	State Test	Free lunch
5	119	0.65	State Test	Free lunch
6	1,573	0.124	Standardized Test	Free lunch
7	1,686	0.175	Standardized Test	Free lunch
8	332	0.54	Standardized Test	Free lunch
9	133	0.06	Standardized Test	Free lunch
10	133	0.43	Standardized Test	Free lunch
11	335	0.166	Standardized Test	Free lunch
12	74	0.43	Achievement Test	Income
13	21,263	0.247	State Test	Income
14	13,279	0.142	State Test	Income
15	415	0.15	Standardized Test	Income
16	120	0.13	GPA	Education
17	302	0.095	GPA	Education
18	696	0.18	GPA	Education
19	113	0.005	GPA	Education
20	3,533	0.34	GPA	Education
21	372	0.06	GPA	Education
22	1,368	0.215	GPA	Education
23	446	0.035	GPA	Education
24	150	0.36	GPA	Education
25	213	0.307	Achievement Test	Education
26	1,328	0.33	Achievement Test	Education
27	1,028	0.16	Achievement Test	Education
28	29	0.334	Achievement Test	Education
29	317	0.403	Standardized Test	Education
30	335	0.202	Standardized Test	Education
31	563	0.18	Standardized Test	Education
32	286	0.23	Standardized Test	Education
33	392	0.44	Standardized Test	Education

Table 11.2 Data for the meta-regression in Table 3.7

Case	Grade	Percent minority	Free lunch	Education level	r	N
1	Primary	0	0	1	1,328	0.33
2	Primary	21	1	0	1,573	0.124
3	Primary	48	0	1	29	0.334
4	Primary	60	1	0	453	0.391
5	Primary	83	0	1	317	0.403
6	Primary	100	0	1	1,028	0.16
7	Elementary	17	0	0	168	0.34

(continued)

Table 11.2   (continued)

Case	Grade	Percent minority	Free lunch	Education level	r	N
8	Elementary	19	1	0	332	0.54
9	Elementary	23	0	1	150	0.36
10	Elementary	36	0	0	143	0.3
11	Elementary	38	0	0	868	0.4
12	Elementary	38	1	0	119	0.65
13	Elementary	55	0	1	392	0.44
14	Elementary	96	0	1	113	0.005
15	Elementary	100	0	1	563	0.18
16	Elementary	100	1	0	133	0.06
17	Elementary	100	1	0	133	0.43
18	Middle	0	0	0	74	0.43
19	Middle	2	0	1	302	0.095
20	Middle	6	1	0	335	0.166
21	Middle	19	0	0	357	0.08
22	Middle	33	1	0	1,686	0.175
23	Middle	49	0	0	398	0.132
24	Middle	75	1	0	85	0.467
25	Middle	100	0	1	120	0.13
26	Middle	100	0	1	286	0.23
27	High school	0	0	0	21,263	0.247
28	High school	0	0	1	3,533	0.34
29	High school	0	0	1	1,368	0.215
30	High school	28	0	1	335	0.202
31	High school	60	0	0	415	0.15
32	High school	76	0	1	696	0.18
33	High school	85	0	0	96	0.01
34	High school	100	0	0	13,279	0.142
35	High school	100	0	1	372	0.06
36	High school	100	0	1	446	0.035
37	Post-secondary	27	1	0	2,307	0.75
38	Post-secondary	31	0	1	1,200	0.315
39	Post-secondary	44	1	0	1,301	0.68
40	Post-secondary	50	0	0	116	0.621

11.2   Hackshaw et al. (1997) Meta-analysis on Exposure to Passive Smoking and Lung Cancer

We use data from Hackshaw et al. (1997) study of the relationship between passive smoking and lung cancer in women to illustrate computations using odds-ratios. The 37 studies included in the meta-analysis compare the number of cases of lung cancer diagnosed in a group of individuals whose spouses smoke with the number of cases of lung cancer diagnosed in individuals whose spouses were non-smokers. Table 11.3 presents the data used in the odds-ratio examples.

Table 11.3 Passive smoking and lung cancer studies

Study	Exposed	Not exposed	Log odds-ratio	Variance
Liu et al.	84	139	−0.301	0.18
Chan et al.	22	133	−0.288	0.08
Kabat et al.	62	190	−0.236	0.339
Wu-Williams et al.	41	196	−0.236	0.016
Buffler et al.	24	25	−0.223	0.193
Brownson et al.	60	144	−0.03	0.013
Lee et al.	134	402	0.03	0.217
Pershagen et al.	29	62	0.03	0.072
Sobue	94	270	0.058	0.034
Shimizu et al.	32	66	0.077	0.071
Kabat et al.	86	136	0.095	0.086
Wang et al.	70	294	0.104	0.066
Sun et al.	20	162	0.148	0.036
Du et al.	199	335	0.174	0.089
Gao et al.	246	375	0.174	0.036
Wu et al.	19	47	0.182	0.231
Garfinkel et al.	54	93	0.207	0.046
Fontham et al.	90	163	0.231	0.01
Akiba et al.	22	47	0.419	0.08
Brownson et al.	90	116	0.419	0.484
Koo et al.	144	731	0.438	0.077
Stockwell et al.	417	602	0.47	0.114
Kalandidi et al.	54	202	0.482	0.09
Lam et al.	23	45	0.501	0.032
Liu et al.	431	1,166	0.507	0.176
Zaridze et al.	210	301	0.507	0.04
Lam	75	128	0.698	0.098
Correa et al.	38	69	0.728	0.227
Trichopolous et al.	651	1,253	0.756	0.089
Geng et al.	67	173	0.77	0.124
Jockel	162	285	0.82	0.317
Humble et al.	230	230	0.85	0.293
Inoue et al.	135	135	0.936	0.398

The table provides the total sample sizes for the group of non-smoking women whose spouses smoked, the group of non-smoking women whose spouses did not smoke, the odds-ratio, and the 95% confidence interval for the odds-ratio. The odds-ratio is the ratio of the odds of being diagnosed with lung cancer given exposure to secondhand smoking to the odds of being diagnosed with lung cancer given no exposure to secondhand smoke. In all but six of the studies, the odds of non-smoking women being diagnosed with lung cancer were higher when they had a spouse who smoked versus non-smoking women whose spouse did not smoke (Table 11.3).

11.3 Eagly et al. (2003) Meta-analysis on Gender Differences in Transformational Leadership

The data in Table 11.4 is adapted from a meta-analysis by Eagly et al. (2003) focusing on gender differences in transformational, transactional and laissez-faire leadership styles. Eagly et al. found that female leaders were more transformational than male leaders, while men tended to use more transactional and laissez-faire types than women. In the examples in the rest of the text using this data, we focus on gender differences in transformational leadership, using characteristics of studies as potential moderators of this gender difference: (a) publication year, (b) average age of the participants, (c) percentage of males in leadership roles in the organization studied, (d) whether the first author is female (1 = female, 0 = male), (e) size of the organization (0 = small, 1 = mixed, 2 = large), (f) whether random selection was used (0 = random, 1 = unsuccessful random, 2 = nonrandom).

Table 11.4 Selected cases from Eagly et al. (2003)

Case	Male N	Female N	Effect size	Variance	Pub year	Age	% male leaders	Female 1st author	Size of org	Random selection
AMA1	963	149	−0.16	0.007761	2001		85	1.00	1.00	1.00
AMA2	613	421	−0.14	0.004016	2001		58	1.00	1.00	2.00
Ay	58	51	−0.19	0.037015	2000			1.00	1.00	1.00
B1	15	8	−0.37	0.194643	1985		65	0.00	2.00	2.00
B2	29	16	−0.24	0.097623	1985		64	0.00	1.00	2.00
B21	574	303	−0.26	0.005081	1996		66	0.00	2.00	2.00
B22	164	107	−0.23	0.015541	1996		60	0.00	1.00	0.00
B23	420	493	−0.09	0.004414	1996	43	46	0.00	1.00	2.00
BJ	112	77	−0.10	0.021942	2000	39		1.00	1.00	1.00
BO	30	31	−0.62	0.068742	1994	50	49	1.00	2.00	1.00
CA	368	240	−0.17	0.006908	1998	38	61	1.00	2.00	1.00
CH	209	111	−0.22	0.013869	1996	45	65	0.00	1.00	1.00
CLS	6,098	2,856	−0.11	0.000515	2000			0.00	1.00	2.00
CU	65	53	−0.17	0.034375	2002	38		1.00	1.00	2.00
CW	456	50	0.61	0.022561	1998	49	90	0.00	2.00	2.00
CW	1,236	132	0.20	0.008399	1999	44	90	0.00	2.00	0.00
DA	27	24	−0.15	0.078924	1996	50		1.00	2.00	2.00
DF	130	72	−0.25	0.021736	1997	49	64	1.00	2.00	2.00
ER	821	699	−0.06	0.002650	1998			1.00	1.00	1.00
EV	16	109	−0.43	0.072414	1997		28	0.00	2.00	0.00
FL	116	77	−0.47	0.022180	1997		90	1.00	1.00	1.00
GM	92	19	−0.10	0.063546	2000	46	83	1.00	2.00	2.00
GO	128	26	−0.25	0.046477	1999		83	1.00	2.00	1.00
HI	29	11	−0.36	0.127012	2000	49	73	1.00	2.00	1.00
JB	134	160	−0.13	0.013741	2000	39	46	0.00	1.00	1.00
JL	288	135	−0.29	0.010979	1996	49	68	1.00	2.00	1.00
JO	6	5	−0.04	0.366739	1992		55	1.00	0.00	0.00

(continued)

Table 11.4 (continued)

Case	Male N	Female N	Effect size	Variance	Pub year	Age	% male leaders	Female 1st author	Size of org	Random selection
KO	296	383	0.31	0.006060	1991	28	42	1.00	2.00	0.00
KP	4,571	1,267	0.04	0.001008	1995		78	0.00	1.00	2.00
KUG	34	7	−0.31	0.173441	1999	35	77	0.00	2.00	1.00
KUU	73	31	0.09	0.045996	1999	55	70	0.00	2.00	1.00
LAN	7	9	−0.26	0.256081	1996			1.00	1.00	2.00
LAV	39	22	−0.33	0.071988	1998		64	1.00	2.00	1.00
LJ	965	288	−0.35	0.004557	1997	59	77	0.00	2.00	1.00
MA	44	59	−0.08	0.039707	2000	43	43	1.00	0.00	1.00
MCG	42	32	−0.72	0.058562	1997	48	57	0.00	2.00	1.00
PO	192	26	−0.27	0.043837	1998		88	1.00	1.00	1.00
RH	229	316	−0.36	0.007650	1996	41	42	1.00	2.00	1.00
RO	29	67	−0.44	0.050416	1993	48	88	1.00	2.00	2.00
SM	112	14	0.00	0.080357	1999	46	85	0.00	2.00	1.00
SP	326	204	−0.21	0.008011	2000	41	62	1.00	2.00	1.00
ST	247	81	0.01	0.016394	2000		74	1.00	0.00	1.00
WH	14	29	0.12	0.106079	2000		33	1.00	2.00	1.00
WI	29	55	−0.87	0.057170	1999	50	35	0.00	2.00	0.00

References

Eagly, A.H., M.C. Johannesen-Schmidt, and M.L. van Engen. 2003. Transformational, transactional, and laissez-faire leadership styles: A meta-analysis comparing women and men. *Psychological Bulletin* 129(4): 569–592.

Hackshaw, A.K., M.R. Law, and N.J. Wald. 1997. The accumulated evidence on lung cancer and environmentaly tobacco smoke. *British Medical Journal* 315(7114): 980–988.

Sirin, S.R. 2005. Socioeconomic status and academic achievement: A meta-analytic review of research. *Review of Educational Research* 75(3): 417–453. doi:10.3102/00346543075003417.

Index

T.D. Pigott, *Advances in Meta-Analysis*, Statistics for Social and Behavioral Sciences,
DOI 10.1007/978-1-4614-2278-5, © Springer Science+Business Media, LLC 2012